Lithium Ion Glassy Electrolytes

Sanjib Bhattacharya · Koyel Bhattacharya
Editors

Lithium Ion Glassy Electrolytes

Properties, Fundamentals, and Applications

Editors
Sanjib Bhattacharya
UGC-HRDC (Physics)
University of North Bengal
Siliguri, West Bengal, India

Koyel Bhattacharya
Department of Physics
Kalipada Ghosh Tarai Mahavidyalaya
Bagdogra, West Bengal, India

ISBN 978-981-19-3271-7 ISBN 978-981-19-3269-4 (eBook)
https://doi.org/10.1007/978-981-19-3269-4

This Springer imprint is published by the registered company Springer Nature Singapore Pte Ltd.
The registered company address is: 152 Beach Road, #21-01/04 Gateway East, Singapore 189721, Singapore

Contents

Part I
Fundamentals of Metal Oxide Glass Composites

Chapter 1
Fundamentals of Lithium-Ion Containing Glassy Systems

Amartya Acharya, Chandan Kr Ghosh, and Sanjib Bhattacharya

Abstract An increasing interest in amorphous solids has grown not only due to their various technological applications in electronic, electrochemical, magnetic, and optical devices, but also from the point of view of their complexity in structure. Glasses are formed as an amorphous (non-crystalline) solid having short-range order; i.e., there is no periodic arrangement of its molecular constituents. The most important aspect of glass transition is the relaxation process that occurs as the supercooled liquid cools. The configurational changes cause the relaxation of the supercooled liquid and become increasingly slow with decreasing temperature, until at a given temperature (glass transition temperature) the material behaves as a solid. Various structural investigations such as SEM, TEM, XRD, FTIR, and FESEM and optical study such as UV–visible have been carried out on different types of glass nanocomposites to explore their various properties.

Keywords Disordered solids–amorphous materials · Glass and glass nanocomposites · Glass transition · Characterization techniques for glass microstructure · Correlation between structure and properties

1.1 Glass

Glass is termed as non-crystalline solid formed by the supercooling of a liquid. **A glass is described as an inorganic fusion product that has cooled to a rigid condition without crystallizing. It is an amorphous solid that exhibits a glass transition** [1]. Amorphous substances are isotropic like gases and liquids and are characterized by short-range order. **According to Turnbull** [2] **and Owen** [3], **glass is a supercooled liquid having the same structure as that of the liquid from which it is frozen. According to McKenzie, any isotopic material, whether inorganic**

A. Acharya · S. Bhattacharya (✉)
UGC-HRDC (Physics), University of North Bengal, Darjeeling, West Bengal 734013, India
e-mail: ddirhrdc@nbu.ac.in

C. K. Ghosh
Dr. B. C. Roy Engineering College, Durgapur, West Bengal, India

S. Bhattacharya and K. Bhattacharya (eds.), *Lithium Ion Glassy Electrolytes*,
https://doi.org/10.1007/978-981-19-3269-4_1

or organic, lacking a three-dimensional atomic periodicity and whose viscosity exceeds about 1014 poise can be described as glass [4]. The main features of a glass in contrast to crystalline compounds are as follows:

1.1.1 It is characterized by a local order that disappears quickly for a long-range disorder giving isotropic property. It is much easier to have a glass with a great homogeneity. "Impurities" are diluted in the matrix "digested" and finally takes place in the framework.

1.1.2 They are characterized by very large compositional possibilities. For example, in the case of oxide glasses, there are many oxide formers such as SiO_2, B_2O_3, V_2O_5, and P_2O_5 and oxide modifiers, Li_2O, CdO, ZnO, and CaO etc. In fact, there are hundreds of thousands of possible glasses.

1.1.3 The verity in the composition is expected to be beneficial for getting various properties of a glassy system such as conductivity, dielectric properties, and mobility of charge carriers. They are characterized by the absence of sharp melting point. The viscosity decreases steadily with a rise in temperature providing possibilities to handle this type of materials. The glass transition (T_g) point may be identified using their high viscous properties (~1013 poise).

1.2 Nanocomposites and Glass Nanocomposites

When in a composite of two or many phases, at least one phase is of the order of nanometer (10^{-9} m) dimension, and the composite material is called a nanocomposite [5]. The precipitation of metals or the formation of cluster in a glass matrix gives birth to glass nanocomposite. The composite thus obtained containing nanoclusters shows different physical properties from those of free atoms and bulk solids having similar chemical composition. Behavior of nanocomposites is sometimes dominated by interphasic interaction and sometimes by the quantum effect associated with the nanostructure. Nanostructured materials are having considerable attention because of their novel physical properties exhibited by matter having nano-dimensional structure [1, 6]. Oxide super-ionic glasses have a random network structure with physical void spaces and, therefore, can be exploited as nano-templates in which distributed nanostructured particles can be generated. These types of oxide glasses containing the distribution of nanoparticles or nanoclusters are designated as glass nanocomposites. The resultant glass nanocomposites exhibit a difference in the electrical conductivity and activation energy as well as mechanical behavior like micro-hardness from that of the host glass matrix.

1.2.1 Classification of Glass Nanocomposites

Assimilation of glass nanocomposite (GNC) properly needs its hierarchical relationships of different types of glasses, which is shown in Fig. 1.1. It shows that glass may be divided into two classes (mainly): inorganic and organic glasses, depending on their chemical and structural composition. Inorganic glasses can be classified into two types depending upon their origin mainly: natural (obsidian or pechsteins) and synthetic (manmade or artificial) glasses. These glasses are used for general purposes in our daily life (e.g., windows, tumblers, bottles, lamps, spectacles, mirrors, etc.) which is collectively known as common glasses.

Metal–glass nanocomposites (MGNCs) consist of nanometal embedded in glass matrix [1, 5, 6]. The nanometal may be silver (Ag), gold (Au), copper (Cu), platinum (Pt), palladium (Pd), etc., or bimetallic alloys (e.g., Ag–Au, Au–Cu, Ag–Pt, etc.) or core–shell nanostructures. The matric glass may persist in the uncrystallized or crystallized state, and the metallic nanostructures may exist encapsulated in uncrystallized and crystallized glassy matrix of the NCs. Similarly, the metallic nanostructures may also be in the amorphous or crystallized state in the MGNCs. The glass is an excellent encapsulating medium owing to its wide range of optical transmission from 0.2 to 20 μm depending on the nature of glassy matrix, which are supposed to be formed by a definite thermodynamic process.

The semiconductor glass nanocomposites (SGNCs) consist of different kinds of semiconductors of the nanometric dimensions as the dispersed phase in different classes of glass matrices [5, 7]. The inorganic semiconductor may be either pure element or the compound. Silicon (Si) and germanium (Ge) are the well-known elemental semiconductors. Compound semiconductors [5, 7] are the compounds of the elements of groups II and VI (e.g., ZnO, ZnS, etc.), groups III and V (e.g., BN, AlN, GaAs, etc.), groups IV and VI (e.g., SnO_2, PbS, PbSe, etc.), and group IV (e.g., SiC, SiGe, etc.). Inorganic semiconductor-doped glasses possess interesting and important optical and electronic properties useful for diverse technologies. Quantum-confined semiconductors or quantum dots (QDs) [5, 8] provide an added dimension to

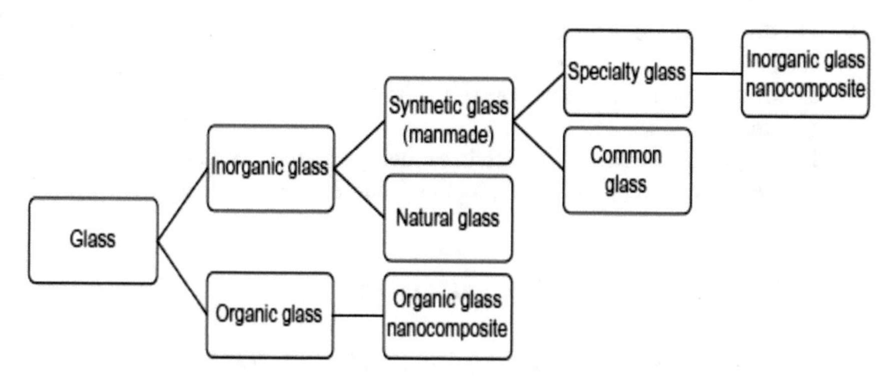

Fig. 1.1 Hierarchical relationships of different types of glasses and glass nanocomposites

the highly active area of "band gap engineering," which deals with the manipulation of the semiconductor band gap and is used for highly efficient and compact solid-state lasers.

Glass–ceramic nanocomposites (GCNCs) are those in which nanocrystals are formed and dispersed in the glassy matrices [5, 9]. Owing to this reason, it may also be regarded as crystal glass nanocomposite (CGNC). The major advantage of these GCNCs is that the nanocrystals can be produced in the glass matrix by applying controlled heat treatment processing. To get it in practice, the temperature and time schedule are to be adjusted very preciously, keeping in mind the information of their compositions, stability, and structures.

The following features of the CGNCs have been found to undergo significant improvements in comparison with that of the conventional composites:

i. Mechanical properties (e.g., modulus, strength, and dimensional stability)
ii. Thermal properties (e.g., conductivity, stability, and heat distortion temperature)
iii. Electrical properties (e.g., conductivity (resistivity) and capacitance)
iv. Optical properties (e.g., absorption or transmission, scattering, and photoluminescence)
v. Chemical properties (e.g., acid–alkali resistance, corrosion, and durability)
vi. Surface properties (e.g., appearance- and surface-enhanced properties).

1.3 Classification of Ionic Glasses

The precipitation of metals or the formation of cluster in a glassy matrix may be the outcomes of the formation of glass nanocomposites [5]. The resultant glass nanocomposites have been shown to exhibit a difference in the electrical conductivity and activation energy from that of the host matrix. The understanding of the structure and the transport properties of glass and glass nanocomposites require the recognition of the following aspects:

(a) Physical structure, which describes the arrangement of atoms with respect to each other.
(b) Chemical structure, which describes the nature of bonding between three different species.
(c) Bonding energy structure, which describes the strength of various bonds.
(d) Electrical properties, that is conductivity, current–voltage characteristics, etc.

1.3.1 Molybdate Glass Nanocomposites

The structure of molybdate glass nanocomposites [10] is derived from several asymmetrical units, mainly MoO_4^{-2} tetrahedral and $Mo_2O_7^{-2}$ ions. Most of the glass nanocomposites containing MoO_3 exhibits absorption peaks near 875, 780, and

320 cm^{-1} (ν1, ν2, and ν3 modes of MoO_4^{-2} tetrahedral ions) which are confirmed from the Fourier transform infrared (FTIR) study. Ionically, conducting glasses and glass nanocomposites containing MoO_3 have attracted much attention because of their potential application in many electrochemical devices such as solid-state batteries, electro-chromic displays, and chemical sensors [11].

1.3.2 Selenite Glass Nanocomposites

The idea of synthesizing selenite glass nanocomposites belongs to Rawson and Stanworth [12] who obtained these in the K_2O–SeO_2 and SeO_2–TeO_2–PbO systems. Dimitriev et al. [13] obtained stable homogeneous glasses with high content of SeO_2 in combination with other non-traditional network formers, viz. V_2O_5, TeO_2, and Bi_2O_3. IR spectra show the independence of SeO_3 pyramids at $\nu^s = 860$–810 cm^{-1} and $\nu^d = 720$–710 cm^{-1}, participated in the network when the SeO_2 concentration is low. As the SeO_2 content increases, SeO_3 groups became associated into chains [14], which contain isolated Se==O bonds with a vibration frequency at 900–880 cm^{-1}.

1.4 Li-Conducting Glasses

For Li^+-conducting glasses, as the radius of Li^+ is small, their mobility is higher than that of other ions because of their easy passage through channels, which makes glass–ceramic an ideal candidate for solid-state electrolyte [15]. Ionic conducting glasses normally consist of a network former, a network modifier, and sometimes a doping salt [16]. Network formers are covalent sulfide or oxides that compose the network [17]. The oxygen or sulfur linking network former polyhedra together is known as bridging oxygen (BO) or bridging sulfur (BS). Network modifiers, when added, break the covalent oxygen or sulfur bridges, creating non-bridging oxygen (NBO) ions or sulfurs (NBS), and each molecule of the modifier introduces mobile ions [18]. Adding network modifiers is expected to decrease the average length of network linkage and makes the glass less rigid, usually decreasing the glass transition temperature [19, 20]. It is predicted that the mechanism of ion conduction in glassy systems involves successive jumps of the lithium-ion between energetically most stable positions near a charge-compensating network pathways [21]. This discussion may conclude that the ionic conductivity in such glassy system depends on the concentration of charge carriers, nature of ions, and mobility of mobile ions.

1.4.1 Brief Review of Some Previous Works

Lithium is considered to be one of the most promising components of rechargeable batteries, which may be a part of various electronic devices such as electric vehicles, smart phones and mobile computers [22, 23]. Lu et al. [24] has pointed out successfully on inconvenience of traditional lithium-ion batteries because of some safety issues, arising due to mixing of highly flammable organic liquid electrolytes or polymer electrolytes. For this reason, traditional organic liquid electrolytes have been replaced by inorganic solid electrolytes that have high thermal stability, high energy density, and better electrochemical stability [25]. Lithium is indispensable to every glass–ceramic, because of its responsibility for the products' zero expansion, ensuring their use in high-temperature ranges without voltage breakage [26]. Effect of silver ion concentration on dielectric properties of Li_2O-doped glassy system [27] has been studied extensively, which reveals remarkable contribution of space charge polarization. Attempts have been made to improve the electrochemical performance of lithium-rich oxide layer material with Mg and La co-doping [28]. Highly resistive lithium-depleted layer [29] has been found in the lithium-conducting composites due to the very low mobile-ion concentration. The interfaces in these nanocomposites have been found to exert a significant control over the ion transport phenomena and other physical properties of materials. All these properties have ensured the Li^+ ion conductor to be developed in glassy forms, as new scopes are revealed in the near future [26]. Relevant works by other researchers reveal the conduction mechanism of other glassy systems, which are as follow.

Gómez-Serrano et al. [30] have shown that the electrical conductivity is strongly dependent on the number of effective electrical contacts established between the sample particles which determine the number of channels or paths available for transport phenomenon of electric current. Under compression, the neighborhood particles are being forced to approach each other which increases the number of electrical contacts and hence the conductivity of the compressed glass nanocomposite [31]. The electrical conductivity of the carbon glass nanocomposites is strongly influenced by the intrinsic conductivity, content, mean crystallite size, and chemical nature of the supported nanoparticles, which is actually dependent on the metal oxide precursor and the tesmperature of heat treatment. In this regard, it is seen that conductivity improves with temperature rise due to the increase in the crystallite size [30].

El-Desoky et al. [32] have observed that generally the maximum enhancement of conductivity is dependent on two reasons:

(i) The increase of concentration of ion pairs during crystallization.
(ii) The grain boundary effect: the numbers of grain boundaries are reciprocal to grain size [33]. It is observed in heat-treated sample that these boundaries are acting as traps to capture electrons and form potential barrier leading to low conductivity. Hence, it is examined that increase in heat treatment time minimizes the number of grain boundaries and electron scattering by increasing the average particle size which lead to the conductivity enhancement [32].

Bhattacharya et al. [34] observed that the DC conductivity of some glass nanocomposite is having dependency on temperature and composition. It follows the Arrhenius law given as $\sigma_{dc} = \sigma_0 \exp(-E_\sigma/k_B T)$, where E_σ is the DC activation energy for glass nanocomposites, T is the absolute temperature, and k is the Boltzmann constant. It happens due to random movement of ion diffusion throughout the composite's network and repeated hops between charge-compensating sites [34]. The AC conductivity, which depends on Almond–West formalism [35–38] given as $\sigma(\omega) = \sigma_{dc}$ $[1 + (\omega/\omega_H)^n]$ where σ_{dc} = DC conductivity, ω_H = hopping frequency, and n = dimensionality, shows a dependency on frequency only at higher frequency region. The composition and temperature dependency are also observed like DC conductivity.

Du et al. [35] have explained that amorphous materials have larger specific surface area with respect to their crystalline counterparts which shows better performance in electrical conductivity and be applicable in more fields. Mostly, the amorphous states consist of less structural confinement when inserting/exchanging large-sized ions than the crystalline counterparts. However, the complexity of understanding of amorphous material is much less than crystal ones as the state can be more easily identified than analyzed. The amorphization of materials also provides an exploration of electrodes for large-size ion storage [37].

Bhattacharya et al. [39] have observed that when a metal oxide glass nanocomposite is doped with other dopant, the electrical conductivity changes according to the nature of dopant. AC conductivity of different doped glassy ceramics shows dispersion at higher frequencies, which is governed by Jonscher's universal power law [38, 40]. This higher frequency conductivity corresponds to a sub-diffusive motion of ions, which strongly depends on inter-ionic interaction and also leads to change in power law exponent of doping oxides. It is also noted that base sample exhibits the higher AC conductivity than the doped one [39].

1.4.2 Applications

- Li_2O–SeO_2–P_2O_5-doped glasses and their nanocomposites might be used as battery materials.
- They may be used as optical sensors.
- The above doped system can be used as cathode materials for Li-ion batteries.

1.5 Mixed Former Effect of Li-Ion-Doped Glassy Systems

A mixed glass former effect has been observed in the Li_2S-doped glassy system [41], which exhibits a large enhancement of ionic conductivity. Here, the electrical conductivity has been correlated with glass transition temperature and the density of the system under study. Structural investigations have been carried out by Raman and SAXS techniques to explore remarkable structural changes, compared to those belonging to the limiting compositions rich in one of the two formers SiS_2 or GeS_2. Raman and SAXS data have pointed toward phase separation for glasses in the central region [41].

The DC conductivity at room temperature and corresponding activation energy for each glassy system are plotted with compositions in Fig. 1.2. Figure 1.3 describes the glass transition temperature (T_g) and density (ρ) of glassy systems with compositions [41]. The estimated values of pre-exponential factor (σ_0) are found to be constant [41]. A sudden break in the curves corresponding to a large increase in conductivity of about 2 orders of magnitude, and a decrease in corresponding activation energy is noteworthy in Fig. 1.2. The change in the modifier Li_2S (30 mol %) also shows similar nature of electrical characteristics for those glasses containing much larger amounts of modifier in the limiting binary systems [41], $yLi_2S - (1 - y)SiS_2$ ($y = 0.5$, $\sigma_{25\,°C} = 1 \times 10^{-4}$ Scm^{-1}, $E_\sigma = 0.32$ eV) and $yLi_2S - (1 - y)GeS_2$ ($y = 0.63$, $\sigma_{25\,°C} = 1.5 \times 10^{-4}$ Scm^{-1}, $E_\sigma = 0.34$ eV).

It is noted in Fig. 1.2 that glass system containing insulating GeS_2 aggregates embedded in a phase close to Li_2SiS_3 might have a conductivity close to the most

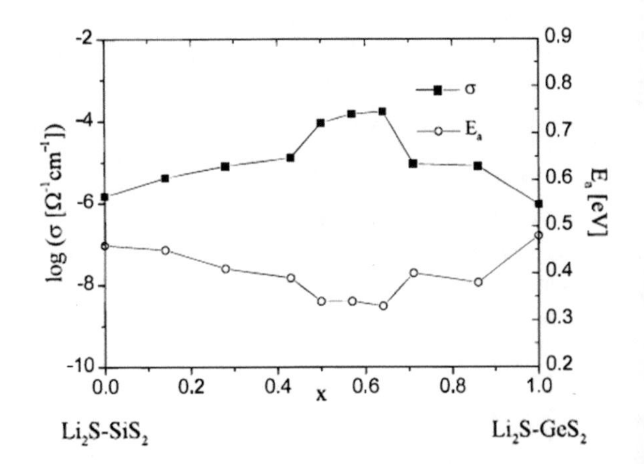

Fig. 1.2 Variations of the conductivity at room temperature $\sigma_{25\,°C}$ and of the activation energy for conduction E_σ with compositions in the glassy system, $0.3Li_2S - 0.7[(1 - x)SiS_2 - xGeS_2]$. Republished with permission from Pradel et al. [41]

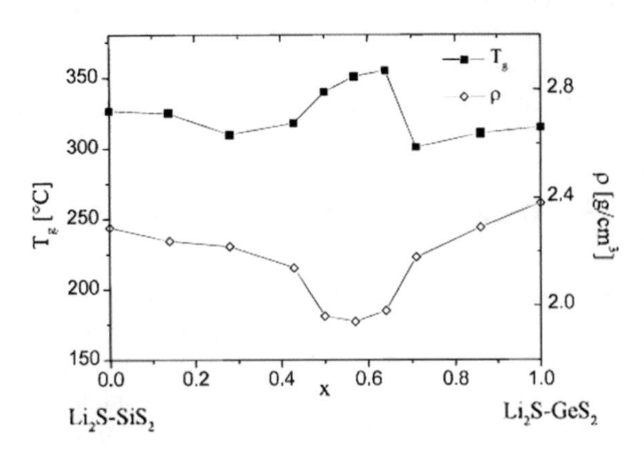

Fig. 1.3 Variations of the glass transition temperature (T_g) and of the density (ρ) with composition for glassy system, $0.3Li_2S - 0.7[(1 - x)SiS_2 - xGeS_2]$. Republished with permission from Pradel et al. [41]

conducting phase, i.e., 10^{-4} Scm^{-1} as it is observed for $0.5 < x < 0.64$, if a percolation threshold is reached [41–43]. In fact, the percolation must have occurred since the volume fraction of each phase is about 0.5 in the central region in Fig. 1.3 as $F(Li_2SiS_3) \approx 2.2$ g cm^{-3} and $F(GeS_2) \approx 2.8$ g cm^{-3} [41]. The strong increase in T_g is also in favor of a strong structural change with appearance of a more rigid phase and does not support the coexistence of the two limiting compositions in the glasses [41]. These results can be closely related to experimental data presented for lithium germanosilicate glasses, which are the oxide counterparts of the thiogermanosilicates under study [41–43]. The germanosilicate glasses are of great interest because of their potential applications in fabrication of optical fibers [41–43]. It is also observed that lithium glasses should be attributed to a phase separation due to the Li^+ tendency to form a Si-based tetrahedra rather than on Ge-based tetrahedral [41–43]. It is also observed [41–43] that one of the resultant glassy "phases" may be the composition of a defined crystalline phase (lithium disilicate). On the other hand, for the system $0.33Li_2O - 0.67[xSiO_2 - (1 - x)GeO_2]$, a composition which corresponds to lithiumdisilicate, homogeneous glasses have been formed with a statistical distribution of Li on Si- or Ge-based tetrahedral [41–43]. In the case of the lithium thiosilicate systems, a crystalline thiodisilicate phase does not exist, but the metathiosilicate Li_2SiS_3 phase does exist [44].

1.6 Key Objectives

- To study the AC conductivity spectra of Li_2O-doped different glassy ceramics in wide frequency and temperature regime
- To study the compositional behavior of the Li_2O-doped glassy ceramics
- To check whether the glassy ceramics are suitable for Li-ion battery application, and how can improvements be done on the properties of the same in near future
- To find a correlation between microstructure and conduction behavior of such glassy system
- To find out thermal and optical properties of such glassy systems.

References

1. S.R. Elliot, *Physics of Amorphous Materials* (England Longman Group Ltd., 1984)
2. D. Turnbull, Contemp. Phys. **10**, 473 (1969)
3. A.E. Owen, Contemp. Phys. **11**, 227 (1970)
4. J.D. Mackenzie, General aspects of the vitreous state, in *Modern Aspects of the Vitreous State*, ed. by J.D. Mackenzie (Butterworth & Co. Publishers Ltd, London, 1960)
5. B. Karmakar, K. Rademann, A. Stepanov, *Glass Nanocomposites: Synthesis, Properties and Applications* (Elsevier B.V., Amsterdam, 2016), ch. 1, pp. 3–53
6. W.H. Dumbaugh, *Corning Glass Works* (Corning, New York, 1985), p. 14831

7. A.C. Wright, *Experimental Techniques of Glass Science* (The American Ceramic Society, Westerville, 1993)
8. B.G. Bagley, *Amorphous and Liquid Semiconductors*, ed. by J. Tauc (Plenum, London, 1974)
9. S.S. Flaschen, A.D. Pearson, W.R. Northover, J. Am. Ceram. Soc. **42**, 450 (1959)
10. S. Bhattacharya, A. Ghosh, J. Appl. Phys. **100**, 114119 (2006)
11. S. Bhattacharya, A. Ghosh, J. Phys. Condens. Matter **17**, 5655 (2005)
12. B. Deb, S. Bhattacharya, A. Ghosh, Europhys. Lett. **96**, 37005 (2011)
13. A.K. Bar, K. Bhattacharya, R. Kundu, D. Roy, S. Bhattacharya, Mater. Chem. Phys. **199**, 322 (2017)
14. S. Bhattacharya, A. Ghosh, Adv. Sci. Lett. **2**, 55 (2009)
15. S. Bhattacharya, A. Acharya, D. Biswas, A.S. Das, L.S. Singh, Phys. B **546**, 10 (2018)
16. S. Bhattacharya, A. Ghosh, J. Chem. Phys. **127**, 194709–194716 (2007)
17. S. Bhattacharya, A. Acharya, A.S. Das, K. Bhattacharya, C.K. Ghosh, J. Alloy. Compd. **786**, 707–716 (2019)
18. S. Bhattacharya, A.K. Bar, D. Roy, J. Adv. Phys. **2**, 241–244 (2013)
19. A. Acharya, K. Bhattacharya, C.K. Ghosh, S. Bhattacharya, Mater. Lett. **265**, 127438–127444 (2020)
20. A. Acharya, K. Bhattacharya, C.K. Ghosh, A.N. Biswas, K. Bhattacharya, Mater. Sci. Eng. B **260**, 114612–114614 (2020)
21. P. Pal, A. Ghosh, J. Appl. Phys. **120**, 045108 (2016)
22. K. Takada, J. Power Sources **394**, 74 (2018)
23. H. Pan, S. Zhang, J. Chen, M. Gao, Y. Liu, T. Zhu, Y. Jiang, Mol. Syst. Des. Eng. **3**, 748 (2018)
24. F. Zheng, M. Kotobuki, S. Song, M.O. Lai, L. Lu, J. Power Sources **389**, 198 (2018)
25. X. Yao, B. Huang, J. Yin, G. Peng, Z. Huang, C. Gao, D. Liu, X. Xu, Chin. Phys. B **25**, 018802 (2016)
26. E. Zanotto, Am. Ceram. Soc. Bull. **89**, 19 (2010)
27. V. Prasad, L. Pavić, A. Moguš-Milanković, A. Siva Sesha Reddy, Y. Gandhi, V. Ravi Kumar, G. Naga Raju, N. Veeraiah, J. Alloys Compd. **773**, 654 (2019)
28. H. Seo, H. Kim, K. Kim, H. Choi, J. Kim, J. Alloys Compd. **782**, 525 (2019)
29. K. Takada, T. Ohno, N. Ohta, T. Ohnishi, Y. Tanaka, ACS Energy Lett. **3**, 98 (2018)
30. J. Sánchez-González, A. Macías-García, M.F. Alexandre Franco, V. Gómez-Serrano, Carbon **43**, 741 (2005)
31. M.M. El-Desoky, M.M. Mostafa, M.S. Ayoub, M.A. Ahmed, J. Mater. Sci. Mater. Electron. **26**, 6793 (2015)
32. A.E. Harby, A.E. Hannora, M.S. Al-Assiri, M.M. El-Desoky, J. Mater. Sci. Mater. Electron. **27**, 8446 (2016)
33. D.P. Almond, G.K. Duncan, A.R. West, Solid State Ion. **8**, 159 (1983)
34. S. Bhattacharya, A. Ghosh, J. Phys. Chem. C **114**(114), 5745 (2010)
35. Z. Wei, D. Wang, X. Yang, C. Wang, G. Chen, F. Du, Adv. Mater. Interfaces **5**, 1800639 (2018)
36. D.P. Almond, A.R. West, Solid State Ion. **9–10**, 277 (1983)
37. X. Yao, B. Huang, J. Yin, G. Peng, Z. Huang, C. Gao, D. Liu, X. Xu, Chin. Phys. B **25**, 018802 (2015)
38. E.M. Assim, E.G. El-Metwally, J. Non-Cryst. Solids **566**, 120892 (2021)
39. S. Bhattacharya, Phys. Lett. A **384**, 126324 (2020)
40. D.P. Almond, A.R. West, Nature **306**, 456 (1983)
41. A. Pradel, C. Rau, D. Bittencourt, P. Armand, E. Philippot, M. Ribes, Chem. Mater. **10**, 2162 (1998)
42. V. Deshpande, A. Pradel, M. Ribes, Mater. Res. Bull. **23**, 379 (1988)
43. A. Pradel, M. Ribes, Mater. Chem. Phys. **23**, 121 (1989)
44. H. Eckert, Z. Zhang, J. Kennedy, J. Non-Cryst. Solids **107**, 271 (1989)

Chapter 2
Lithium-Ion-Doped Glassy System

Koyel Bhattacharya and Sanjib Bhattacharya

Abstract An increasing interest in lithium-ion-doped glassy system has grown not only due to their various technological applications in electronics, solid-state batteries, sensors, optical devices, etc., but also from the point of view of academic interest. The conductivity spectra of such glassy systems have been studied in a wide frequency range from a few hertz to few terahertz. The nature of such conductivity spectra has been analyzed by considering a high-frequency vibrational motion as well as the hopping motion of the mobile ions in the DC conductivity plateau regime, the dispersive regime, and the high-frequency plateau regime, respectively. The redox stability of lithium-doped glassy systems has been determined by cyclic voltammetry and by Wagner polarization method. In case of Li_2O-doped oxide glassy system containing transition metal ions, Li^+ are in the intercalated structure in such a way that Li^+ is dispersed in the voids of the glassy matrix in association with localized electron in the *d* orbitals of the transition metals.

Keywords Lithium-ion-doped glassy system · High-frequency (THz) regime · DC conductivity plateau regime · Dispersive regime · Li_2O-doped and other oxide glassy systems

2.1 Introduction

A solid-state electrolyte is a key material for constructing a solid-state battery, which can have higher electrical energy density without the leakage problem of liquid electrolytes [1]. New materials hold the key to fundamental advances in energy conversion and storage, both of which are vital in order to meet the challenge of global warming and the finite nature of fossil fuels [2]. Lithium batteries are the systems

K. Bhattacharya
Department of Physics, Kalipada Ghosh Tarai Mahavidyalaya, Bagdogra, Darjeeling, West Bengal 734014, India
e-mail: koyel21stapril@gmail.com

S. Bhattacharya (✉)
UGC-HRDC (Physics), University of North Bengal, Darjeeling, West Bengal 734013, India
e-mail: ddirhrdc@nbu.ac.in; sanjib_ssp@yahoo.co.in

S. Bhattacharya and K. Bhattacharya (eds.), *Lithium Ion Glassy Electrolytes*,
https://doi.org/10.1007/978-981-19-3269-4_2

of choice, offering high energy density, flexible, lightweight design and longer life span than comparable battery technologies [3]. Since metallic lithium combines with an anode having negative redox potential and low equivalent weight. This anode combined with a cathode electrode material and rechargeable cell can be constructed [4]. Recent literature survey [5] reveals that the traditional Li^+ ion batteries have been found to have issues relating to safety, due to organic liquid electrolytes, which are highly flammable. To avoid these issues, inorganic-solid electrolytes are used as replacement, due to their higher energy density and higher electrochemical and thermal stability [6]. The interfaces in these nanocomposites have been found to exert a significant control over the ion transport phenomena and other physical properties of materials. All these properties have ensured the Li^+ ion-conductor to be developed in glassy forms, as new scopes are revealed in the near future [7].

2.2 Lithium-Ion-Doped Glassy System (Various Systems)

Cramer et al. [8] studied and presented conductivity spectra of various fast ion-conducting glassy system containing Li_2O and halides. The conductivity spectra of these glassy systems [8] were studied in a wide frequency range from a few hertz to few terahertz. In their research [8], the nature of conductivity was analyzed into two distinguished directions, one was due to a high-frequency vibrational motion of mobile ions, and the other was due to the hopping motion of the mobile ions. The latter part consists of three regimes [8]: the DC conductivity plateau regime, the dispersive regime, and the high-frequency plateau regime. In their study [8], DC regime exhibited the dependency of the conductivity on glass composition. The complete spectra of the hopping motion was interpreted using "unified site relaxation model (USRM)" [8], which was based on the assumption that different types of ionic site in the glassy matrix were existing. In USRM [8], complete hop in the conductivity spectra was formed by a superposition of two types of contributions and the concept of mismatch and relaxation [9] were considered to be promising mechanism. Backward-hop tendency [8, 9] might be more pronounced in these glassy systems with higher connectivity of the respective components [10, 11].

Ghosh et al. [12] presented the observations with the assimilation of the essential microscopic parameters and their roles in the macroscopic ion dynamics in different glassy systems such as CdI_2–$LiPO_3$ and LiI–$LiPO_3$. The results of the characteristic length scales followed a direct correlation with the ionic conductivity [12]. They have beautifully shown the roles of the slight modifications of the glassy networks in the composition dependence of microscopic parameters. In their systematic examination on ion dynamics [12] in several single and mixed former glassy systems in wide composition and temperature ranges, they have successfully shown strong dependency of ionic conductivity of them on the dopant salt content as well as on the mixed former ratio. The characteristic lengths such as mean square displacement and spatial extent of sub-diffusive motion of lithium ions are supposed to be the key parameter to explain ion dynamics of such glassy systems. The correlation between both, ionic conductivity and the characteristic lengths with the modification of such glassy network structure, are the prime novelty of this work.

Otto [13] reported glassy system with high Li^+ conductivities at relatively low temperatures in the year 1966. Various compositions in the system, M_2O–B_2O_2–SiO_2 (M = Li, Na, K) were reported by him [13], and still, they are considered to be the most promising compositions for Li^+ conductors not only for various applications but also for academic interest. Li^+-conducting borate glasses with 40 wt% lithium compound [13] exhibited highest conductivity (10^{-1} Ω^{-1} cm^{-1}) at 350 °C. It is interestingly noted that activation energy of electrical conduction increases linearly with increasing Li_2O content in the glassy system containing less than 25 wt% lithium compound [13]. It is also noted that replacing up to 50 wt% B_2O_3 by SiO_2 shows a small change in the value of activation energy for a given alkali content [13]. Lithium content of the glasses is further increased by incorporating lithium salts like Li_2SO_4, LiCI or LiF, and this helps in obtaining enhanced conductivities than in those containing Li_2O alone. Tuller et al. [14] later verified the results of conductivity spectra by introducing complex impedance plots. These results on the borate system inducted interest to many researchers to study on crystalline fast ion-conducting compositions with boracite structure [15, 16] even in the glassy form. In this regard, the works by Levasseur et al. [17, 18] are noteworthy. Ionic conductivity of such glassy system [17, 18] is found to be a strong function of polarizability of the anion, which may directly indicate that the conductivity must be a function of ionic radius of the anion. The high conductivity of such glassy system [14–18] has been suggested to be due to the salt-like structure with higher Li_2X levels. Here, glass transition temperature shows dependence on heating and cooling rate during its measurement and also on the wt% of lithium compounds in the present glassy system.

2.3 Advantage and Disadvantage Such Glassy System

The total experimental conductivity spectrum of particular Li_2O-doped glassy system [8, 9] has been presented in Fig. 2.1 at a particular temperature. Hop and vibration are clearly mentioned in Fig. 2.1. Hop conductivity [8, 9] is usually pronounced in terms of a high-frequency plateau, which is alike to those known to exist in crystalline ion conductors such as $RbAg_4I_5$ [10] and Na-β-alumina [11]. In Fig. 2.1, the high-frequency plateaus [8, 9] are not appeared in the hop-frequency extremity due to the presence of broad vibrational peaks superimposing the high-frequency part of hop conductivity. In the vibrational frequency range [8, 9], the applied field is supposed to be changed its sign rapidly in such a manner that every hop or displacement may contribute to the conductivity.

Ghosh et al. [12] studied structural vibrations form FT-IR spectra of different compositions of the LiI–$LiPO_3$ glassy system as shown in Fig. 2.2. Here, the intensity of the dominant P=O bond is found to decrease with LiI content due to the shortening of the phosphate chains, which may be the consequence of the increase

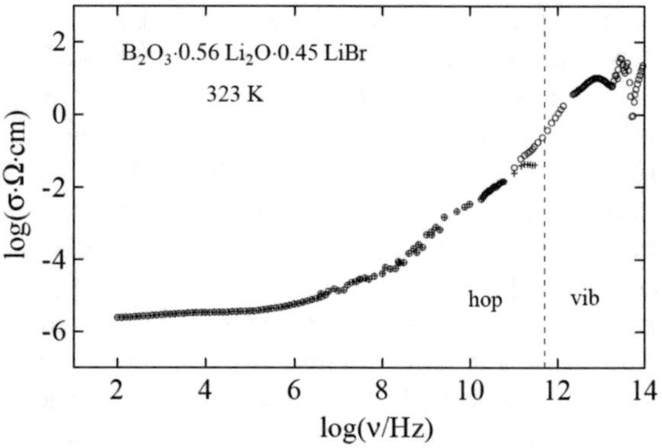

Fig. 2.1 Total experimental conductivity spectrum of glassy $B_2O_3{\cdot}0.56\ Li_2O{\cdot}0.45$ LiBr at 323 K (O) and spectrum obtained after removing vibrational contributions out of it (+). Republished with permission from Funke and Cramer [8]

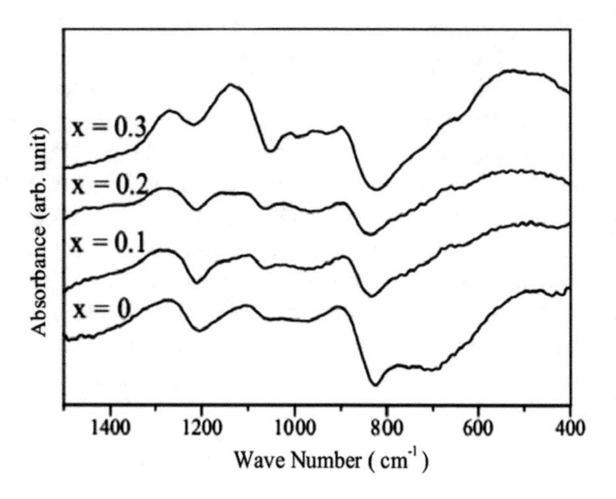

Fig. 2.2 FT-IR spectra for different compositions of xLiI − (1 − x)$LiPO_3$ glassy system. Republished with permission from Shaw et al. [12]

of pyrophosphate ($P_2O_7^{4-}$) groups. It is also mentioned in this literature [12] that the electron density of the P=O bond in $P_2O_7^{4-}$ groups is redistributed to the oxygen at the chain end, which is may play important role decrease the intensity of P=O vibration. As the band of P–O$^-$ modes increases slightly with the increase of LiI content, the phosphate network is depolymerized due to the creation of the non-bridging oxygen with the incorporation of LiI into $LiPO_3$ glassy system [12]. They have correlated the depolymerization process which has been correlated with microscopic parameters [10] for the present glassy system. But it needs more structural studies such as atomic absorption spectroscopy, Raman and FT-Raman spectroscopy, and positron annihilation spectroscopy to validate such results.

The redox stability of lithium-doped glassy systems has been determined by cyclic voltammetry and by Wagner polarization method [19]. A typical voltagram obtained for such glassy system is depicted in Fig. 2.3. The cathodic and anodic limits are characterized, respectively, by a reversible peak at $e_o = -1.9$ V and by a non-apparent reversible peak at $e_o = +1$ V. The cathodic peak may be considered as a reduction of

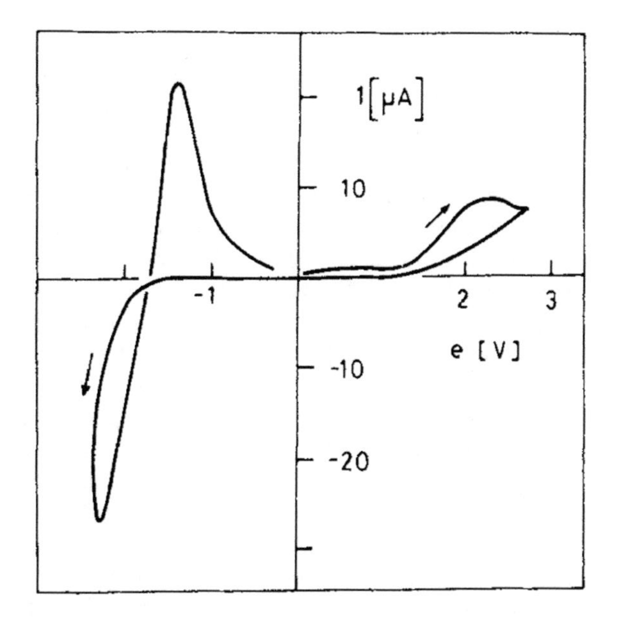

Fig. 2.3 Cyclic voltammetry curve for a LiI–Li_2S–P_2S_5 glass [19]. Republished with permission from Kulkarni et al. (1984)

Li^+, while on the return cycle, the oxidation peak may correspond to the dissolution of 80% of the deposited metal. In the anodic range, the electrochemical reaction relative to the oxidation wave is attributed to the sulfides strongly bonded to the glassy network. The discharge characteristics of a cell using LiI–P_2S_s–Li_2S glassy system has explored [19] at room temperature. An output curve obtained for the cell Li/glass/WO_3 is presented in Fig. 2.4 [19]. It may be noted that the discharge at 25 °C at a current density of 20 pA shows the cell stability for nearly 4 h. But more works are required to explore electrochemical stability of such glassy systems. Use of lithium foil should have a chance to be oxidized, which needs globe box for performing this experiment. This arrangement makes this process quite costly.

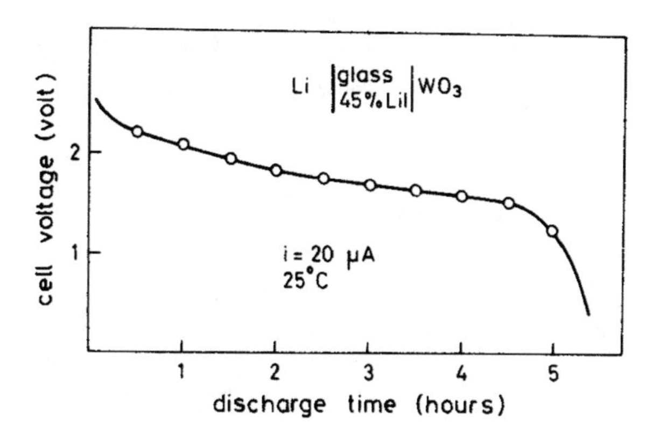

Fig. 2.4 Discharge characteristics of a LiI–Li_2S-P_2S_5 in Li/Glass/WO_3 cell [19]. Republished with permission from Kulkarni et al. (1984)

2.4 Comparison of Lithium-Doped Glassy Systems with Other Oxide Glassy Systems

The term “oxide glassy system” means traditional soda-lime silicate glassy system having network structures, made of SiO_4 tetrahedral building blocks connected by mixed ionic-covalent bonding [20]. In search of various oxide glassy systems, metallic glasses are most important because of their icosahedral structural units with metallic bonding [20]. Organic polymeric glasses are made through the cross-linking of molecular chains involving van der Waals bonding between chains and covalent intra-molecular bonding [20]. Metal–organic framework glasses are newly developed with connected tetrahedral via coordination bonds [20].

Transition metal ions doped oxide glassy systems [21] are known to exhibit electronic/polaron conductivity. As the transition metal oxides do not easily form a glass alone, the stability of glassy phases can be improved by the addition of a conventional network former such as SiO_2, P_2O_5, or TeO_2 [21]. In case of Li_2O-doped oxide glassy system containing transition metal ions, Li^+ are in the intercalated structure in such a way that Li^+ are dispersed in the voids of the glassy matrix in association with localized electron in the *d* orbitals of the transition metals [22]. But electron displacement may be interpreted due to disordered structures as per scheme of small-polaron model [21–23]. In some commercial purposes [24], SnO is reduced to metallic tin (Sn) and lithium forms Li_2O during the first discharge in a Li_2O–B_2O_3–P_2O_5 glassy system in such a manner that tin (Sn) reacts with further lithium to the composition limit to develop a specific anodic capacity of 991 vmAh g^{-1}.

Literature survey [25] reveals that the nature of the modifier cations is equally important as that of the former because the modifier may be considered as the major compositional component due to its cross-linking motif. Correlation between structures and electrical transport and ionic interactions cannot only be found in invert glasses (silicates, phosphates, and borates with more modifier than former), but also in other types of oxide glassy systems such as oxoanionic (nitrates, sulfates, carbonates, and hydrates), halides (fluoride, chloride, bromide, and iodide), super-ionic (fast ion conductors or solid electrolytes, the specific conductivity (σ) is usually within the range from about 10^{-3} to 10 Ω^{-1} cm^{-1}) silver phosphates and lithium sulfides and combinations of these glass families [26–28].

2.5 Conclusion

Lithium-ion-doped glassy systems and their various aspects related to electrical, structural, and optical properties have been extensively discussed. The conductivity spectra of such glassy systems have been analyzed in the directions of high-frequency vibrational motion as well as the hopping motion of the mobile ions. The later phenomena have been interpreted in three distinguished regions, namely DC conductivity plateau regime, the dispersive regime, and the high-frequency plateau regime, respectively. The redox stability of lithium-doped glassy systems has been determined by cyclic voltammetry and by Wagner polarization method. Li_2O-doped oxide glassy system containing transition metal ions are supposed to be formed with an idea of Li^+ intercalated structure for dispersion of Li^+ in the voids of the glassy matrix.

References

1. Y. Ogiwara, K. Echigo, M. Hanaya, J. Non-Cryst. Solids **352**, 5192 (2006)
2. C.A. Angell, Solid State Ion. **105**, 15 (1998)
3. A.K. Singh, Opt.-Int. J. Light Electron Opt. **124**, 2187 (2013)
4. G. Rishab, R. Rohit, K. Deepak, IJSTE **3**, 114 (2016)
5. V. Kumari, A. Kaswan, D. Patidar, N. Saxena, K. Sharma, Process. Appl. Ceram. **9**, 61 (2015)
6. D.D.H. Singh, K.S. Rathore, N. Saxena, J. Asian Ceram. Soc. **6**, 1 (2018)
7. M. Sevi, S. Asokan, Phys. Rev. B **58**, 4449 (1998)
8. K. Funke, C. Cramer, Curr. Opin. Solid State Chem. Mater. Sci. **2**, 483 (1997)
9. K. Funke, B. Roling, M. Lange, Solid State Ion. **105**, 195 (1998)
10. R. Hoppe, T. Kloidt, K. Funke, Ber. Bunsenges. Phys. Chem. **95**, 1025 (1991)
11. K.L. Ngai, U. Strom, Phys. Rev. B **38**, 10350 (1988)
12. A. Shaw, B. Deb, S. Kabi, A. Ghosh, J. Electroceram. **34**, 20 (2015)
13. K. Otto, Phys. Chem. Glass. **7**, 29 (1966)
14. H.L. Tuller, D.P. Button, D.R. Uhlmann, J. Non Cryst. Solids **40**, 93 (1980)
15. W. Jeitschko, T.A. Bither, Z. Naturforsch. B **27**, 1423 (1972)
16. W. Jeitschko, T.A. Bither, P.E. Bierstedt, Acta Crystallogr. B **33**, 2767 (1977)
17. A. Levasseur, B. Cales, J.M. Reau, P. Hagenmuller, Mater. Res. Bull **13**, 205 (1978)
18. A. Levasseur, B. Cales, J.M. Reau, P. Hagenmuller, Mater. Res. Bull. **14**, 921 (1979)
19. A.R. Kulkarni, H.S. Maiti, A. Paul, Bull. Mater. Sci. **6**, 201 (1984)
20. C.R. Cao, K.Q. Huang, J.A. Shi, D.N. Zheng, W.H. Wang, L. Gu, H.Y. Bai, Nat. Commun. **10**, 1 (2019)
21. S. Bhattacharya, A. Ghosh, J. Appl. Phys. **100**, 114119 (2006)
22. N.F. Mott, E.A. Davis, *Electronic Processes in Non-Crystalline Materials* (Oxford University Press, 1971)
23. I.G. Austin, N.F. Mott, Adv. Phys. **18**, 41 (1969)
24. Y. Idota, T. Kubota, A. Matsufuji, Y. Maekawa, T. Miyasaka, Science **276**, 1395 (1997)
25. H.J.L. Trap, J.M. Stevels, Glastech. Ber. **5**, 32 (1959)
26. A.C.M. Rodrigues, M.L.F. Nascimento, C.B. Bragatto, J.L. Souquet, J. Chem. Phys. **135**, 1 (2011)
27. M. Tatsumisago, K. Hirai, T. Hirata, M. Takahashi, T. Minami, Solid State Ion. **86–88**, 487 (1996)
28. M. Tatsumisago, T. Minami, M. Tanaka, Yogyo Kyokaishi **93**, 581 (1985)

Chapter 3
Methods of Preparation of Lithium Ion-Doped Glassy Systems

Koyel Bhattacharya and Sanjib Bhattacharya

Abstract Li_2O-doped glassy amorphous systems are supposed to be less thermodynamically stable than the corresponding crystalline form (i.e., possesses greater free energy). Preparation of such glassy materials may be considered by adding excess free energy in some way to the crystalline polymorph. There are extensive selections of methods for the synthesis of such glassy systems like melt-quenching, sol–gel, chemical vapor deposition, sputtering, etc. Out of these, the melt-quenching technique and sol–gel process are very widespread, simple, and very easy to prepare different types of glassy materials, particularly on the laboratory scale.

Keywords Melt-quenching · Sol–gel · Chemical vapor deposition · Sputtering

3.1 Introduction

Li_2O-doped glassy systems and their nanocomposites are of significant attention in consequence of their microstructural features and distinctive properties [1, 2]. These materials are also subjects of great interest as they contain the characteristic features of disordered (amorphous) materials and some properties of crystalline materials. In fact, structure, composition, and the nature of the bonds of such glassy system and nanocomposites usually control the electrical properties [3, 4]. The slight variation in the bond length and bond angles disturbs the spatial periodicity of the structure [5] as they reveal short-range order rather than long-range order [5]. Studies on the spectroscopic properties, for instance, optical absorption, infrared spectra, and structural properties could be utilized as probes to elucidate the structural aspects of such glassy materials. The spectroscopic studies may facilitate for the explanation of the electrical conduction mechanism of the present glass nanocomposite materials.

K. Bhattacharya
Department of Physics, Kalipada Ghosh Tarai Mahavidyalaya, Bagdogra, Darjeeling, West Bengal 734014, India
e-mail: koyel21april@gmail.com

S. Bhattacharya (✉)
UGC-HRDC (Physics), University of North Bengal, Darjeeling, West Bengal 734013, India
e-mail: ddirhrdc@nbu.ac.in; sanjib_ssp@yahoo.co.in

S. Bhattacharya and K. Bhattacharya (eds.), *Lithium Ion Glassy Electrolytes*,
https://doi.org/10.1007/978-981-19-3269-4_3

The prerequisite of classifications of the materials is to decide the amorphousness of the glassy nanocomposite samples with the occurrence of certain crystallinity. Unless one has the above structural information, a reasonable understanding of the electrical and dielectric behavior of the glassy systems under consideration would have been rather challenging. X-ray powder diffraction (XRD) technique is an indispensable method of material exploration and characterization as XRD has many outstanding features and advantages that promote its extensive range use [6, 7]. XRD is a process employed to examine the crystallinity of the glassy systems, i.e., recognition of the crystalline phases, the spacing between lattice planes, percentage crystallinity of each phase, and the sizes of crystallites. Since every material has its exclusive diffraction patterns, therefore, materials or compounds can be predicted considering the database of diffraction patterns. It has several applications to industrial and academic research that eventually enrich the growth and development of science and technology. Ultraviolet–visible spectroscopy (UV–Vis) describes absorption spectroscopy or reflection spectroscopy into nanocomposite samples in the ultraviolet–visible spectral range [8, 9]. In the presence of UV–Vis electromagnetic spectrum, atoms and molecules adopt electronic transitions from the ground state to the excited state. The UV–Vis absorption spectrum can be utilized to compute the optical band gap energy of the semiconducting materials (allowed direct, allowed indirect, forbidden direct, and forbidden indirect transitions). Using UV–Vis spectroscopy of a compound, it is possible to find out Urbach Energy, which indicates the degree of defect states present in the compound. Fourier transform infrared spectroscopy (FT-IR) is a method for observing an infrared spectra of the absorption into a solid, liquid, or gas [10, 11]. A FT-IR spectrometer simultaneously accumulates data with high spectral resolution over an extended spectral range. The term Fourier transform infrared spectroscopy is stand on the point that Fourier transform is mandatory to convert the experimental data into the actual spectrum. Fourier transform infrared spectroscopy is utilized to categorize the existence of different bonds and binding transformations. Field emission scanning electron microscopy (FE-SEM) is an investigative technique to find very small topographic evidence in the surface, entire, or fractioned sample materials [12–14]. Executing this technique, researchers usually identify the structure of materials that may be as small as 10 nm. FE-SEM technique is applied to yield real space-magnified images of a surface displaying what it looks like. In general, FE-SEM microscopy refers to surface crystallography (i.e., how atoms are arranged on the surface), surface morphology (i.e., the form and size of topographic features that make up the surface), and surface composition (the elements and compounds from which the surface consists). Transmission electron microscopy (TEM) is a microscopic method wherein an electron beam passes through a sample to form an image instead of light used in a normal microscope [15, 16]. TEM uses high-energy electrons to find morphological, composite, and crystallographic information of a sample, at a maximum magnification of 1 nm. TEM provides two-dimensional high-resolution images that allow an extensive range of educational, scientific, and industrial applications. Positron annihilation spectroscopy (PAS) is a non-destructive spectroscopic technique for investigating the existence of micro-voids and defects in amorphous or crystalline materials [17–19]. The coincidence Doppler broadening spectroscopic (CDBS) measurements establish the transformation of surroundings around the positron trapping defects. It also confirms the presence of vacancy-type defects arising from cation non-stoichiometry and the amorphous character of the glassy systems.

3.2 Various Methods of Preparations of Lithium Ion-Doped Glassy Systems

As the amorphous phase is less thermodynamically stable than the corresponding crystalline form (i.e., it possesses greater free energy), the preparation of Li_2O-doped amorphous glassy materials can be regarded as the addition of excess free energy in some manner to the crystalline polymorph. Faster the rate of cooling (or deposition), further the amorphous material/solid lies from equilibrium. For a long time, it was thought that only a relatively restricted number of materials could be prepared in the form of amorphous solids, and it was common to refer to these 'special' substances (e.g., oxide glasses and organic polymers) as ***'glass forming solids'***. It is now realized that ***'glass forming ability'*** is almost a universal property of condensable matter. The correct viewpoint is: ***Nearly all materials can, if cooled fast enough and far enough, be prepared as amorphous solids***.

There are several methods by which the Li_2O-doped glassy materials/solids can be prepared. Only the most important and widely followed methods are discussed here.

3.2.1 *Melt-Quenching Followed by Heat Treatment*

The oldest and one of the most popular and established methods is the quenching of melt. In this process, an amorphous material is cooled from the molten phase very quickly. The amorphous materials formed in this manner are often termed as 'glasses'; as they exhibit glass transition phenomenon. The distinguishing feature of the melt-quenching process of producing amorphous materials is that the amorphous solid is formed by the 'continuous' hardening (i.e., increase in viscosity) of the melt. In contrast, crystallization of the melt occurs as a discontinuous solidification, solid growth taking place only at the liquid–solid interface, with the result that crystallites grow in the body of the melt.

The essential prerequisite for 'glass' formation from the melt, therefore, is that the cooling be sufficiently fast to preclude crystal nucleation and growth. The crystalline phase is thermodynamically more stable, and the crystal growth will always dominate over the formation of the amorphous phase. The condition for glass formation is that the nucleation rate should be less than a certain value, 10^{-6} cm^{-1} s^{-1}.

The most rapid melt-quench technique is the 'splat-quenching method', developed specially for metallic glass proposed by Duwez et al. [20] which achieved the cooling rate of 10^5–10^8 K.

When the glasses are passed through a heat treatment above their glass transition temperatures, different nanophases are found to grow in the host glass matrix. Tatsumisago et al. [21] prepared the glass samples $70Li_2S–30P_2S_5$ and heated them at 240 °C for 2 h (glass transition temperature was found to be 210 °C) to obtain glass–ceramic samples. They found that the conductivity of the heat-treated samples was much greater than their glassy counterparts. The conductivity enhancement was assumed to be due to the precipitations of highly lithium ion-conducting crystals from the $Li_2S–P_2S_5$ glasses.

3.2.2 Gel Desiccation

The technique for producing amorphous materials via the sol–gel process has considerable technological promise [22]. The method has its greatest usefulness for those systems which give rise to very viscous melts near the melting point (and consequently have considerable difficulty in achieving homogeneous mixing), or alternatively which have extremely high melting points and hence pose considerable technical problems in actually being able to make a glass by melt-quenching.

Amorphous materials are prepared via a gel, using either aqueous solution or organic materials as the starting components:

- Destabilization of a sol (commonly of silica), with other components being added in the form of appropriate aqueous solutions.
- Hydrolysis and polymerization of mixture of organometallic compounds.

In both methods, however, a multi-component 'gel' (the elastic solid product produced abruptly from a viscous liquid by process of continuing polymerization), being non-crystalline and homogeneous, is heated to remove volatile components and cause an initial densification, which is completed by a final process of sintering or fusion to produce the amorphous solid.

The method employing aqueous components starts with a silica 'sol' of silicic acid, $Si(OH)_4$ (a dispersion at the molecular level in aqueous solution) to which are added other components (e.g., metal salts in aqueous solution). The gelling process is influenced by several factors, e.g., pH of the medium, particle size, and temperature. The gel can be converted to an amorphous solid in one of three ways: heating at temperatures below the glass transition temperature, T_g, resulting in chemical polymerization; sintering (with or without the application of pressure) at temperature above T_g but much below the melting point; fusing the gel particulate to form a glass.

3.2.3 Thermal Evaporation

This technique is perhaps conceptually the most easy to understand and is possibly the most widely used method for producing amorphous thin films. It is one of several ways of producing amorphous solids by deposition from a vapor. In this technique, the starting compound is vaporized and the material is collected on a substrate. In general, this process is performed in vacuum. The starting compound is taken in a powdered form. It is heated either by taking it in a 'boat' and passing a high dc current (for low melting point compounds) or by hitting with the high-energy electron beam (for high melting point compounds). This process depends upon several factors, some of which can be controlled by the experimenter, but others can vary from one preparation to other. The substrate temperature is one of them. If the substrate temperature is suitably high, then it can lead to higher surface mobility which produces materials with considerably fewer structural defects. However, if the substrate temperature is too high, the material will crystallize.

3.2.4 Sputtering

This process consists of the bombardment by energetic ions from a low-pressure plasma, causing erosion of material, either atom by atom or as clusters of atoms, and subsequent deposition of a film on the substrate. Sputtering is significantly superior to evaporation for the production of multi-component systems. As the sputtering rates do not vary widely for different components, so for a multi-component target, the sputtered films tend to preserve the stoichiometry of the starting material.

A further option is offered by sputtering in the use of gas other than Ar which chemically reacts with the target, resulting in 'reactive sputtering'. This can significantly increase the sputtering rate, as well as incorporate chosen additives into the films.

3.2.5 Chemical Route

This approach involves selecting suitable precursor chemicals, which was subjected to heat treatment under different atmospheric conditions. Huber et al. [23] used the pressure injection technique for preparing nanowire arrays of metals (In, Sn, and Al) and semiconductors (Se, Te, GaSb, and Bi_2Te_3) within the channels of anodic alumina membranes. A novel method for surfactant-assisted growth of crystalline copper sulfide nanowires was reported by Wang et al. [24].

3.2.6 Template Assisted Growth

The electrodeposition inside nanoporous membrane templates has provided a versatile approach to prepare nanowires of metals, semiconductors, and polymers [25]. In comparison to other procedures, template methods are generally inexpensive, allowing deposition of a wide range of nanowire materials and presenting the ability to create very thin wires (5 nm–10 μm), with aspect ratios (length over diameter) as high as 1000 [26, 27]. Another important advantage of the template method is that the nanowires can be diameter-controllable and well defined (i.e., the template provides an effective control over the uniformity, dimensions, and shape). In particular, porous anodic aluminum oxide (AAO) has nanometer-size channels (about 5–250 nm in diameter), with high pore densities (up to 1011 pores/cm^2), and controllable channel lengths (a few nanometers to hundreds of micrometers). For this reason, AAO can serve as an ideal template for the growth of monodispersive nanowires [28, 29]. The diameter and density of the pores are controlled by varying the anodization conditions of high-purity aluminum. In addition, the AAO templates are especially suited for use at higher temperatures (thermally stable up to 1000 °C) [29].

3.2.7 Other Techniques

Beside the above described techniques, there are many other techniques for the preparation of amorphous materials.

'Glow-discharge decomposition' technique is such a technique for the preparation of amorphous thin films. In this technique, amorphous thin film can be produced by glow-discharge decomposition in the vapor phase. This technique, like sputtering, relies on the production of a plasma in a low-pressure gas, but instead of ions from the plasma ejecting (sputtering) materials from a target, chemical decomposition of the gas itself takes place, leading to deposition of a solid film on a substrate placed in the plasma. The plasma is produced by application of an r.f. field.

Another remarkable technique for the preparation of amorphous system is 'chemical vapor deposition'. This technique is quite similar to glow-discharge method. The difference is that the chemical vapor deposition process relies on thermal energy for the decomposition, and the applied r.f. field (if used) simply serves to heat up the substrate upon which the vapor decomposed. Temperatures of the order of 1000 K are commonly used.

There are also a few other techniques like electrolytic deposition, irradiation, shock-wave transformation, etc. through which amorphous materials can be prepared in bulk or thin film form.

3.3 Some Advantages and Disadvantages of Various Methods (Cost-Effective, Usefulness, etc.)

Indeed, the vast majority of glasses are prepared by the melt-quenching method [3]. In this process, the batch of the assortment of weighed reagent grade materials and the assortment melts in the temperature ranges contingent on the composition. After casting, the liquefied is properly equilibrated near the glass transition temperature of the respective glasses to remove the residual thermal stresses. After that, the melt suddenly cools down to room temperature, and liquefied have been quenched quickly between two aluminum plates. The obtained glasses are called the as-prepared glass nanocomposites. The batch is prepared by proper selection of the raw materials followed by chemical calculation, weighing, and mixing. The homogenization of the glass melt is confirmed by intermittent agitation of the molten mass by a platinum or silica glass rod. The glass-melting crucible may be made of fused silica (SiO_2) or alumina (Al_2O_3). Sometimes, twin-roller process is used to get thin flake-shaped samples instead of quenching of melts by aluminum plates due to the fact that formation of glassy samples may require lower amount of exposed surface. The technique for preparing lithium glassy materials (amorphous) via the sol–gel process has the considerable technological promise [22]. The method is very useful for making very viscous melts near the melting point. But, it has considerable difficulty in achieving homogeneous mixing. The gel can be converted to an amorphous solid in one of three ways: heating at temperatures below the glass transition temperature (T_g) resulting in chemical polymerization; sintering (with or without the application of pressure) at temperatures above T_g but much below the melting point; fusing the gel particulate to form a glass. The electrodeposition inside nanoporous membrane templates has provided an adaptable methodology preparing nanophases of metals, semiconductors, and polymers [25, 29]. In comparison to other procedures, template-assisted growth methods are usually cheap, allowing deposition of a wide range of nanowire materials and offering the ability to create very thin wires (5 nm–10 μm) with aspect ratios (length over diameter) as high as 1000 [26, 27]. Another remarkable advantage of the template method provides effective control over uniformity, size, and shape of various nanophases. In particular, porous anode alumina has nanometer channels (diameter about 5–250 nm), high pore density (up to 1011 pores/cm^2), and controllable channel lengths (several nanometers to hundreds of micrometers). For this reason, porous anode alumina can serve as an ideal template for the growth of monodisperse nanoelement [28, 29]. But, porous anode alumina templates are particularly suitable for use at higher temperatures (heat stable up to 1000 °C) [29]. Melt-quenching technique and sol–gel process are very widespread and cost-effective methods.

3.4 Conclusion

The development of Li_2O-doped glassy systems and their various preparation route are discussed. In this regards, melt-quenching, sol–gel, chemical vapor deposition, sputtering, thermal evaporation, template-assisted growth, and other processes have been mentioned. Out of these, the melt-quenching technique and sol–gel process are very widespread, simple, and very easy to prepare different types of glassy materials, particularly on the laboratory scale. Some advantages and disadvantages of various methods have also been discussed.

References

1. K. Takada, J. Power Sources **394**, 74 (2018)
2. H. Pan, S. Zhang, J. Chen, M. Gao, Y. Liu, T. Zhu, Y. Jiang, Mol. Syst. Des. Eng. **3**, 748 (2018)
3. B. Karmakar, K. Rademann, A. Stepanov, *Glass Nanocomposites: Synthesis, Properties and Applications* (Elsevier B.V., Amsterdam, 2016)
4. F. Zheng, M. Kotobuki, S. Song, M.O. Lai, L. Lu, J. Power Sources **389**, 198 (2018)
5. X. Yao, B. Huang, J. Yin, G. Peng, Z. Huang, C. Gao, D. Liu, X. Xu, Chin. Phys. B **25**, 018802 (2016)
6. J.A. Bearden, X-ray wavelengths and X-ray atomic energy levels. Rev. Mod. Phys. **31**(1) (1967)
7. P. Bertin Eugene, *Introduction to X-Ray Spectrometric Analysis* (Springer Science & Business Media, 2013), pp. 1–485. ISBN 9781489922045
8. H.-H. Perkampus, *UV-VIS Spectroscopy and Its Applications* (Springer-Verlag Berlin Heidelberg, 1992), pp. 1–244
9. B.M. Tissue, Ultraviolet and visible absorption spectroscopy, in *Characterization of Materials* (2012), pp. 1–13
10. P. Griffths, J.A. De Haseth, *Fourier Transform Infrared Spectrometry*, 2nd edn. (Wiley, 2007), pp. 1–16
11. R. Bhargava, Infrared spectroscopic imaging: the next generation. Appl. Spectrosc. **66**, 1091 (2012)
12. N.L. Donovan, G.H. Chandler, S. Supapan, *Scanning Electron Microscopy* (Wiley, 2002)
13. P.J. Goodhew, J. Humphreys, R. Beanland, *Electron Microscopy and Analysis*, 3rd edn. (Taylor & Francis, 2001), pp. 1–236
14. T.L. Kirk, *Near Field Emission Scanning Electron Microscopy* (Logos Verlag Berlin, 2010), pp. 1–97. ISBN-13: 9783832525187
15. D.B. Williams, C. Barry Carter, *Transmission Electron Microscopy: A Textbook for Materials Science* (Springer, 2012), pp. 3–89. ISBN: 978-0-387-76500-6
16. C. Carter Barry, B. Williams David, *Transmission Electron Microscopy Diffraction, Imaging, and Spectrometry* (Springer, 2011), pp. 1–166
17. J.W. Humberston, M. Charlton, *Positron Physics* (Cambridge University Press, 2001)
18. P. Hautojärvi, A. Dupasquier, *Positrons in Solids* (Springer-Verlag, 1979)
19. R. Krause-Rehberg, H.S. Leipner, *Positron Annihilation in Semiconductors*. Solid-State Sciences (Springer-Verlag, Berlin, 1999)
20. P. Duwez, R.H. Willens, R.C. Crewdson, J. Appl. Phys. **36**, 2267 (1965)
21. F. Mizuno, A. Hayashi, K. Tadanaga, M. Tatsumisago, Adv. Mater. **17**, 918 (2005)
22. C.J. Brinker, G.W. Scherer, *Sol–Gel Science: The Physics and Chemistry of Sol-Gel Processing* (Academic Press Inc., San Diego, 1990)
23. C.A. Huber, T.E. Huber, M. Sadoqi, J.A. Lubin, S. Manalis, C.B. Prater, Science **263**, 800 (1994)

24. S. Wang, S. Yang, Chem. Phys. Lett. **322**, 567 (2000)
25. T.L. Wade, J.E. Wegrowe, Eur. Phys. J. Appl. Phys. **29**, 3 (2005)
26. A.J. Yin, J. Li, W. Jian, A.J. Bennett, J.M. Xu, Appl. Phys. Lett. **79**, 1039 (2001)
27. G. Meng, A. Cao, J.Y. Cheng, A. Vijayaraghavan, Y.J. Jung, M. Shima, P.M. Ajayan, J. Appl. Phys. **97**, 064303 (2005)
28. N.C. Seeman, Nature (Lond.) **421**, 427 (2003)
29. J.D. Le, Y. Pinto, N.C. Seeman, K. Musier-Forsyth, T.A. Taton, R.A. Kiehl, Nano Lett. **4**, 2343 (2004)

Chapter 4
Features of Lithium-Ion Doped Glassy Systems

Sanjib Bhattacharya

Abstract Glass formation principle is discussed using rapid quenching method. Thermal and elastic properties of some lithium-doped glassy systems have been discussed and analyzed. Structural basis, mainly molybdate and selenite, has been mentioned with suitable glassy systems to explore their roles in glass formation processes. Technological applications of various lithium-doped glass nanocomposites have been discussed.

Keywords Glass formation principle · Melt-quenching route · Structural basis · Technological applications

4.1 Introduction

In general, lithium oxide-doped glassy systems are supposed to be non-crystalline solids (NCS) with lacking of long range positional order [1, 2]. Customarily, such a network, is considered as a set of vertices (representing centers of atoms) connected by strong short-range (i.e., covalent) bonds so that a path of bonds exists between any two vertices. According to this view, weakly bonded systems do not possess network structure [3]. Networks can be crystalline (having translational periodicity which is a special form of positional long-range order), quasi-crystalline (having long-range positional order without translational periodicity) or non-crystalline (lacking long-range positional order). For most solids, the crystalline state is the natural one since the energy of the ordered atomic arrangement is lower than that of an irregular packing of atoms [3]. However, when the atoms are not given an opportunity to arrange themselves properly, by inhibiting their mobility, amorphous materials may be formed; an example is amorphous carbon formed as a decomposition product at low temperatures. Certain polymers are composed of very large and irregular molecules, and in such cases, a crystalline packing is not easily obtained [3]. In other cases, the solid state may correspond to a supercooled liquid in which the

S. Bhattacharya (✉)
UGC-HRDC (Physics), University of North Bengal, Darjeeling, West Bengal 734013, India
e-mail: ddirhrdc@nbu.ac.in; sanjib_ssp@yahoo.co.in

S. Bhattacharya and K. Bhattacharya (eds.), *Lithium Ion Glassy Electrolytes*,
https://doi.org/10.1007/978-981-19-3269-4_4

molecular arrangement of the liquid state is frozen in; because of rapid cooling and high viscosity of liquid, crystals may not have had enough time to grow and a glassy material results [3]. Randomness can occur in several forms, of which topological, spin, substitutional, and vibrational disorders are the most important [3]. Disorder is not a unique property; it must be compared to some standard and that standard is the perfect crystal [1, 2].

An increasing interest in amorphous solids has grown not only due to their various technological applications in electronic, electrochemical, magnetic, and optical devices [1–6], but also from the point of view of their complexity in structure. The study of amorphous materials started much later, compared to crystalline materials and in spite of a large number of investigations already made [1, 2, 7, 8], the glass formation ability and physical properties of these materials are not well understood. This is because of the complicated theories and models needed to explain the non-periodic potential to the electrons in the amorphous materials, in sharp contrast to the well-known band theory for the crystalline materials [9].

From the technological point of view, the advantages of the amorphous materials over their crystalline counterparts are manifold. Firstly, large area homogeneous thin films of amorphous nature are easy to prepare. Nowadays, a-Si semiconductors are in huge commercial use as solar cells, photosensors, flat screen displays, etc. Secondly, bulk glasses can be readily formed from the melt by slow quenching and the materials remain workable (i.e., the viscosity is relatively low) over a range of temperatures. This particular property allows the materials to be easily fashioned into various shapes, specifically drawn into long thin fibers, which has been the key issue for the recent progress in optical communication [1, 2]. Moreover, as a result of their structural homogeneity, the physical properties of the amorphous materials are isotropic, unlike crystalline materials for which the intrinsic behavior of single crystal may be anisotropic and the presence of grain boundaries in polycrystalline samples may dominate the overall behavior. Furthermore, amorphous phases can be formed in mixed—component systems over wide range of compositions, which allows their properties to continuously vary with composition [1, 2].

The study of amorphous materials started much latter, compared to crystalline materials, and in spite of a large number of investigations already made [8, 9], the glass formation ability and physical properties of these materials are not well understood. This is because of the complicated theories and models needed to explain the non-periodic potential to the electrons in the amorphous materials, in sharp contrast to the well-known band theory for the crystalline materials [9].

4.2 Glass Formation Principles

In general, a glass is neither a liquid nor a solid; rather, it has a distinctly different structure with properties of both liquids and solids. It would be convenient if one could conclude that glassy materials change from being a supercooled liquid to an amorphous solid at the "***glass transition temperature***." Thus, one can define glass more generally as, "***glasses are amorphous materials, which exhibit glass transition phenomenon***."

The glass transition is a phenomenon, in which a solid amorphous phase exhibits a more or less abrupt change in derivative thermodynamic properties (such as specific heat and thermal expansion coefficient) from "crystal-like" to "liquid-like" values with the change in temperature. This definition has an advantage that the term glassy is confined to those materials, which can be obtained in a reproducible state, since the materials can be in a state of internal equilibrium above the glass transition. These changes can be observed readily by monitoring the volume as a function of temperature (using a dilatometer). A typical result is shown in Fig. 4.1. It can be observed that the liquid ↔ crystal transformation is characterized by an abrupt change of slope in volume at the melting or freezing temperature (T_f).

On the other hand, the liquid ↔ glass transformation exhibits a gradual break in slope and the region over which such change of slope occurs is termed as "glass transition temperature" T_g. As the glass transition temperature is not well defined, another temperature called the fictive temperature, which is obtained by the intersection of the extrapolated liquid and glass curves is defined.

The nature of glass transition is very complex and is even now poorly understood. Many attempts have been made toward its understanding [1, 2, 8, 9]. During glass transition, both the specific heat and thermal expansion coefficient change in a narrow temperature range from a low-value characteristic of crystal to a high-value characteristic of liquid. Thus, from the thermodynamic aspects of glass transition this behavior is very close to that expected for a second-order phase transition. It is worthwhile to mention that a second-order phase transition involves a discontinuity in the specific heat, heat capacity, etc., which are the second order of the Gibbs function. However, these changes for glasses are not as sharp as they should be in a true second-order phase transition, but instead are diffuse occurring over a small temperature interval rather than at a sharply defined temperature. This is also the case

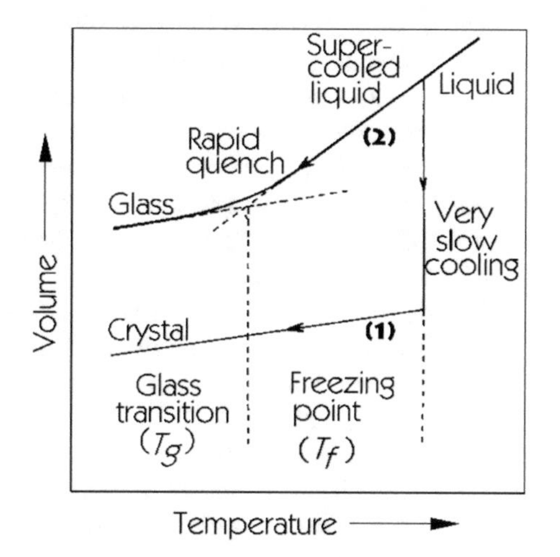

Fig. 4.1 Schematic illustration of the two cooling paths by which liquid may solidify. A very slow cooling rate leads to a discontinuous change in volume to a crystal state (curve 1). Rapids quench leads to a continuous change in volume (curve 2)

for other thermo-dynamical variables such as entropy and enthalpy. This implies that at T_g there should be a discontinuity in derivative variables, such as coefficient of thermal expansion $\alpha_T = (\partial \ln V/\partial T)_P$, compressibility $\kappa_T = (\partial \ln V/\partial P)_T$, and heat capacity $C_P = (\partial H/\partial T)_P$.

The kinetic aspects of the glass transition are also important. The glass transition temperature depends on cooling rate. This influence of the cooling rate on glass transition is the clearest proof that the glass transition differs from a strict thermodynamic transition. The dependence of the glass transition temperature, understood in terms of interplay between the time scale of the experiment and the kinetic or molecular recovery, is the main manifestation of the kinetic dimension of the glass transition. However, there are other aspects, which bear materially on the question of an underlying thermodynamic phase transition.

The most important aspect of glass transition is the relaxation process that occurs as the supercooled liquid cools. The configurational changes cause the relaxation of the supercooled liquid and become increasingly slow with decreasing temperature, until at a given temperature (glass transition temperature) the material behaves as a solid. For time of observation long compared with the structural relaxation time, the material appears liquid-like, while for time of observation shorter than structural relaxation time the material behaves a solid-like. A transition takes place if the values of liquid-like parameters differ significantly from solid-like ones. Thus, a glass transition occurs, when the time of observation is equal to the structural relaxation time. For temperatures below glass transition temperature, the structure tends to approach the equilibrium state of the supercooled liquid. This process can occur in times of the order of minutes for temperature near glass transition temperature, but may take years for temperature far below glass transition temperature.

Another theory which concerns with certain aspects of glass transition and have many similarity with aspects of relaxation theory is the free volume theory. In this theory, total volume of a liquid is supposed to be divided into two parts: One part is occupied by the atoms or molecules, and the other part provides free space for them to move. The latter volume permitting diffusive motion is termed free volume. As the temperature of a liquid is lowered, both the occupied volume and the free volume are expected to contract. The glass transition occurs, when the free volume of the supercooled state decreases below some critical value. The redistribution of free volume no longer occurs; i.e., the free volume is frozen-in in the locations when the glass is formed. When percolation aspects are taken into consideration, the free volume theory predicts the glass transition to be more likely a first-order transition in contrast to thermodynamic theory.

A study of the thermal properties of $(70 - y)B_2O_3 - 30BaF_2 - yLiX$, where X = F, Cl, Br, indicated that initially ($y \leq 10$ mol%) the halide adds to the network forming B–X–Li bonds with smaller-radii halides being more likely to add to the network, and then after $y \geq 10$ mol%, halide atoms add interstitially as per study of El-Hofy and Hager [10]. This result is manifested in the variation of glass transition temperature with LiX as shown in Fig. 4.2 [11].

Figure 4.3 shows the variation of elastic moduli with LiX [11]. Surprisingly, all lithium halide salts are found to increase in density and elastic moduli [11]. In this

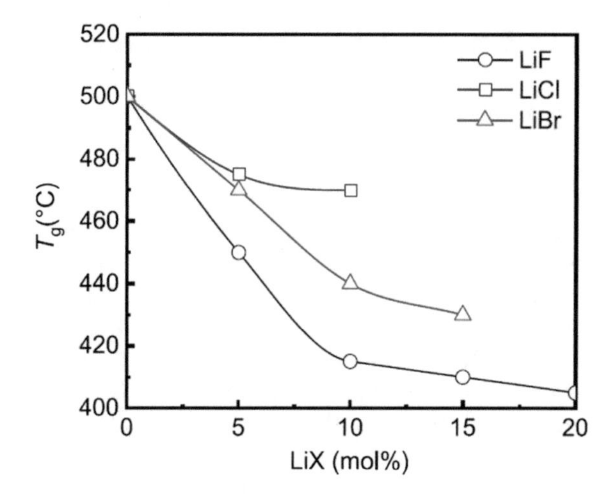

Fig. 4.2 Glass transition temperature T_g as a function of lithium halide content in $(70 - y)B_2O_3 - 30BaF_2 - yLiX$ for X = F, Cl, Br. Republished with permission from Calahoo and Wondraczek [12]

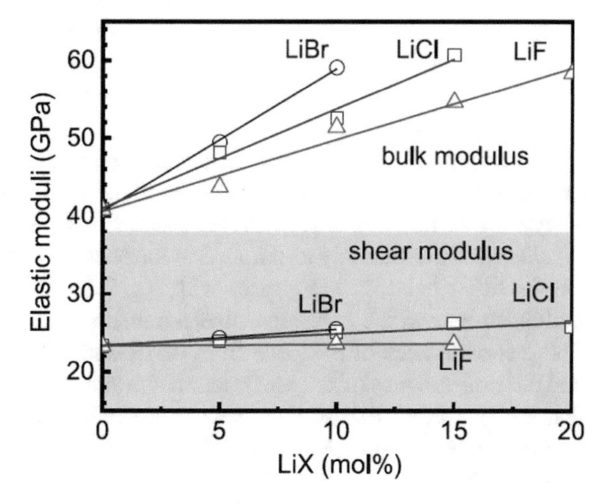

Fig. 4.3 Variation of elastic moduli with lithium halide content in $(70 - y)B_2O_3 - 30BaF_2 - yLiX$, where X = F, Cl, Br. Republished with permission from Calahoo and Wondraczek [12]

enhancement process, it is also interestingly noted that that chlorine and bromine should play vital role in increasing the elastic moduli more than fluorine, which can explained by differences in bond length, the stretching force constant between the bonds and the average cross-linking density [13, 14].

4.3 Structural Basis for Glass Formation

4.3.1 Molybdate Basis

There are various structural basis/units of lithium-doped glassy systems and nanocomposites. Here, only two basis molybate and selenite are discussed. In some cases, other alkali-doped iodite glassy systems are included in this section, which is expected to be similar as LiI-doped glassy systems.

The structure of molybdate glasses is constructed from several asymmetric units, mainly MoO_4^{2-} tetrahedral and $Mo_2O_7^{2-}$ ions [15]. Most of the glasses and glass nanocomposites containing MoO_3 exhibits absorption peaks 875, 780, and 320 cm^{-1} (ν_1, ν_2, and ν_3 modes of MoO_4^{2-} tetrahedral ions) which are confirmed from the FT-IR study [15, 16].

Kawamura et al. [17] showed that the progressive change of activation energy observed in the molybdate glasses could be attributed to the order–disorder transition in the α-AgI crystal. The ionic conductivity of these glasses occurred due to the cooperative liquid-like motion of the mobile ions and the network structure of glasses probably cause the non-Arrhenius behavior in the rapidly quenched molybdate glasses.

Eckert et al. [18] demonstrated the local structures of molybdenum species in the glassy system using near infrared Fourier transform (NIR-FT) Raman spectroscopy. They showed that in glasses with the Ag_2O/MoO_3 ratio of unity, the molybdenum species were present only as tetrahedral monomeric orthomolybdate ions, MoO_4^{2-}. On the other hand, in the glasses with Ag_2O/MoO_3 molar ratios less than unity, molybdenum species were present as tetrahedral orthomolybdate anions, MoO_4^{2-}. The preponderance of evidence from NMR and vibrational spectroscopy suggests that this unit contains linked MoO_4 tetrahedra and MoO_6 octahedra. They also showed that the structure of these units was probably similar to the chain ions present in crystalline $Na_2Mo_2O_7$. Minami and Tanaka [15] showed that glasses with molar ratio $Ag_2O/MoO_3 = 1$ contained no condensed macroanions, but only discrete Ag^+, I^-, and MoO_4^{2-}. In their model, only a part of the silver ions were believed to participate in the conduction process.

Recent report [19] on electrical properties of semiconducting tellurium molybdate glasses had adequately been explained using small polaron theory. This report also showed that the glass-forming oxide greatly affected the magnitude of the conductivity and the activation energy for hopping conduction.

Ionically conducting glasses and glass nanocomposites containing MoO_3 have attracted much attention because of their potential application in many electrochemical devices such as solid-state batteries, electrochromic displays, and chemical sensors [20]. In particular, silver ion-conducting glasses are at the focus of current interest, because of their high stability against humidity and their high electrical conductivity in the range of 10^{-1} S/cm at room temperature. Glassy system containing MoO_3, first reported by Minami [15, 16], belong to this group of materials, and their glass-forming regions, electrical properties, glass-transition temperatures, and local structures have been examined extensively [15, 16, 20–23].

The glass-forming region of the superionic system containing MoO_3 [18] and the structure of such glassy systems have been investigated by many researchers using IR [15, 16, 24], Raman [25], EXAFS [26], and neutron diffraction [27]. While many studies agree that the molybdenum species exist as tetrahedral orthomolybdate anion MoO_4^{2-} in these glasses [15, 16, 25, 26], and other reports have claimed octahedral molybdenum environment [27, 28]. Typical structural features in crystals with the dimolybdate stoichiometry have also been explored by a group of researchers [18, 24, 25]. Comparatively, few studies have been carried out on glasses with composition ratio $Ag_2O/MoO_3 < 1$. In these glasses, two additional infrared absorption bands have been observed at 600 and 450 cm^{-1} [15, 16]. The structure of crystalline $Na_2Mo_2O_7$ [29] is based on infinite chains formed by MoO_4 tetrahedra and MoO_6 octahedra [30]. An interpretation of the IR spectra has been given by Caillet et al. [31]. Inspection of the crystal structures of other crystalline molybdates based on $Mo_2O_7^{2-}$ anionic units reveals a substantial structural variety. Infinite chains of interlinked MoO_4 and MoO_6 units also exist in the compounds $K_2Mo_2O_7$ [32] and $(NH_4)_2Mo_2O_7$ [33]. Discrete dimeric bitetrahedral $Mo_2O_7^{2-}$ anions are known to be present in (n-$Bu_4N)_2Mo_2O_7$ [34], $(PPN)_2Mo_2O_7$ ($PPN = [Ph_3P = N = PPh_3]^+$) [35], $MgMo_2O_7$ [36], $K_2Mo_2O_7$–KBr double salt [37], $K_2Mo_2O_7$ melt [38], and (n-$Bu_4N)_2Mo_2O_7$–CH_3CN solution [34]. The crystal structure of $Ag_2Mo_2O_7$ consists of infinite chains formed by blocks of four edge-shared MoO_6 octahedra joined by edge-sharing [39]. In principle, all of these arrangements provide possible explanations for the extra IR bands observed in glasses with Ag_2O/MoO_3 ratios < 1. Molybdate glasses and their nanocomposites are particularly interesting because of the growing evidence of anomalous in the structure as well as the intensive properties when compared with silver borate and silver phosphate glasses [39, 40]. Experimentally, considerable efforts have been made to establish a relation between the microscopic structure and fast ion conductivity in glasses and glass-nanocomposites.

4.3.2 Selenite Basis

The idea of synthesizing selenite glasses belongs to Rawson [41] and Stanworth [42] who obtained glasses in the K_2O–SeO_2 and SeO_2–TeO_2–PbO systems. Dimitriev et al. [43] obtained stable homogeneous glasses with high content of SeO_2 in combination with other non-traditional network formers: V_2O_5, TeO_2, Bi_2O_3. They showed from the IR spectra the independent SeO_3 pyramids $\nu^s = 860 - 810$ cm^{-1} and $\nu^d = 720 - 710$ cm^{-1} participated in the network when the SeO_2 concentration was low. As the SeO_2 content was increased, SeO_3 groups became associated into chains which contain isolated Se=O bonds with a vibration frequency at $900 - 880$ cm^{-1}.

Satyanarayana et al. [44] studied the differential scanning calorimetry (DSC) for the AgI–Ag_2O–SeO_2–V_2O_5 glasses. They observed a decrease in T_g with the increase in AgI content. They suggested that a larger number of bonds were destroyed within the glassy network in order to allow its rearrangement to form a more open-type thermodynamically stable phase.

Venkateswarlu et al. [45] showed that the dc conductivity of $AgI–Ag_2O–SeO_2–V_2O_5$ system increases with the AgI content and exhibited highest conductivity ($\sigma = 2.63 \times 10^{-2}\ \Omega^{-1}\ cm^{-1}$) for the $66.67\%AgI–23.07\%Ag_2O–10.26\%(0.8SeO_2 + 0.2V_2O_5)$ glass system with further increase of AgI content the conductivity was found to decrease. This type of decrement of conductivity was explained from their structural behavior.

4.4 Steps of Manufacturing Such Glassy Systems

The most accepted route for manufacturing glassy system is the melt-quenching route [46]. In this route, the molten form of the material is cooled quickly to stop the crystal growth [46]. Quenching rates play a significant role in the preparation of glassy solids.

Many materials need sufficiently rapid quenching in order that the melt solidifies into glass. Rates of cooling required for glassy phase formation are different for different materials. In this technique, the chemicals in the definite proportions are mixed and melted and the melt is quickly quenched to room temperature. Normally, the crystalline phase is thermodynamically more stable than the amorphous phase. In order to form glassy state, it is necessary to maintain the rate of cooling of the melt to be sufficiently fast enough to prevent crystallization. When the glasses are passed through a heat treatment above their glass transition temperatures, different nanophases are found to grow in the host glass matrix. Tatsumisago et al. [14] prepared the glass samples $70Li_2S–30P_2S_5$ and heated them at 240 °C for 2 h (glass transition temperature was found to be 210 °C) to obtain glass–ceramic samples. They found that the conductivity of the heat-treated samples was much greater than their glassy counterparts. The conductivity enhancement was assumed to be due to the precipitations of highly lithium-ion-conducting crystals from the $Li_2S–P_2S_5$ glasses.

4.5 Technological Approaches

Lithium is considered to be one of the most promising components of rechargeable batteries, which may be applicable to electric vehicles, smart phones and mobile computers [47–49]. Some safety issues have been pointed out [49, 50], which may arise due to mixing of highly flammable electrolytes. For this reason, traditional electrolytes have been substituted by inorganic solid electrolytes with enough thermal stability, energy density and electrochemical stability [49, 51, 52]. Influence of silver ion concentration on dielectric properties of Li_2O doped glassy system [49] has been studied extensively, which reveals remarkable contribution of space charge polarization. Attempts have been made to enhance the electrochemical performance of lithium-rich oxide layer material with Mg and La co-doping [49]. Highly resistive lithium-depleted layer [49] has been found in the lithium-conducting composites due to the very low mobile-ion concentration. Conductivity spectra [53] have been studied by many researchers. Besides these ionic systems containing Li^+, polaron-hopping in semiconducting lithium-doped glass nanocomposites containing transition metal ions (TMI) are also very much interesting not only for academic interest but also for the application in optical switching, display devices, etc. As such, a thorough investigation of the electrical transport in both ion-conducting and semiconducting lithium-doped glasses and glass nanocomposites will be significant for the better understanding of the conduction and relaxation processes in these technical materials.

4.6 Conclusion

Slow and fast cooling processes for glass formation principle are discussed using graphical analysis. Nature of various lithium halides has been discussed to explore their thermal and elastic properties. Researchers have developed various structural basis such as molybdates and selenites to form various lithium glassy systems and to explore their roles in glass formation processes. Both ion-conducting and semiconducting lithium-doped glasses and glass nanocomposites are of great interest for the better understanding of the conduction and relaxation processes in these technical materials.

References

1. S.R. Elliot, *Physics of Amorphous Materials*, 2nd edn. (Longman Group UK Limited, 1990)
2. R. Zallen, *Physics of Amorphous Solids* (Wiley, New York, 1983); Torquato S, Nature **405**, 521 (2000)
3. P. Hagenmuller, W. Van Gool (eds.), *Solid Electrolytes: General Principles, Characterization, Materials, Applications* (Academic Press, New York, 1978)
4. S. Chandra (ed.), *Superionic Solids, Principles and Applications* (North Holland, Amsterdam, 1981)
5. E.C. Subbarao (ed.), *Solid Electrolytes and Their Applications* (Plenum Press, 1980)
6. B.V.R. Chowdari, S. Radhakrishna (eds.), *Materials for Solid State Batteries* (World Scientific, Singapore, 1986)
7. P.G. Le Comber, J. Mort (eds.), *Electronic and Structural Properties of Amorphous Semiconductors* (Academic, London, 1973)
8. W. Vogel, *Chemistry of Glass*, ed. by N. Kreidl (American Ceramic Society, Ohio, 1985)
9. N.F. Mott, E.A. Davis, *Electronic Process in Non-Crystalline Materials* (Clarendon, Oxford, 1979)
10. M. El-Hofy, I.Z. Hager, Phys. Status Solidi Appl. Res. **199**, 448 (2003)
11. I.Z. Hager, M. El-Hofy, Phys. Status Solidi Appl. Res. **198**, 7 (2003)
12. C. Calahoo, L. Wondraczek, J Non-Cryst. Solids: X. **8**, 100054 (2020)
13. K. Funke, R.D. Banhatti, Mater. Res. Soc. Symp. Proc. **756**, 81 (2003)

14. F. Mizuno, A. Hayashi, K. Tadanaga, M. Tatsumisago, Adv. Mater. **17**, 918 (2005)
15. T. Minami, M. Tanaka, J. Non-Cryst. Solids **38–39**, 289 (1980)
16. T. Minami, T. Katsuda, M. Tanaka, J. Non-Cryst. Solids **29**, 389 (1978)
17. N. Kuwata, T. Saito, M. Tatsumisago, T. Minami, J. Kawamura, Solid State Ion. **175**, 679 (2004)
18. N. Machida, H. Eckert, Solid State Ion. **107**, 255 (1998)
19. A. Ghosh, Phil. Mag. **61**, 87 (1990)
20. A. Sanson, F. Rocca, G. Dalba, P. Fornasini, R. Grisenti, New J. Phys. **9**, 88 (2007)
21. P. Mustarelli, C. Tomasi, E. Quartarone, A. Magistris, Phys. Rev. B **58**, 9054 (2003)
22. A. Sanson, F. Rocca, P. Fornasini, G. Dalba, R. Grisenti, A. Mandanici, Phil. Mag. **87**, 769 (2007)
23. P. Ghigna, M.D. Muri, P. Mustarelli, C. Tomasi, A. Magistris, Solid State Ion. **136–137**, 479 (2000)
24. E.I. Kamitsos, J.A. Kapouotsis, G.D. Chryssikos, J.M. Hutchinson, A.J. Pappin, M.D. Ingram, J.A. Duffy, Phys. Chem. Glass. **36**, 141 (1995)
25. S. Muto, T. Suemoto, M. Ishigame, Solid State Ion. **35**, 307 (1989)
26. K.J. Rao, J. Wong, S. Hemlata, Proc. Indian Acad. Sci. (Chem. Sci.) **94**, 449 (1985)
27. K. Suzuki, J. Non-Cryst. Solids **192/193**, 1 (1995)
28. A. Rajalakshmi, M. Seshasayee, G. Aravamudan, T. Yamaguchi, M. Nomura, H. Ohtaki, J. Phys. Soc. Jpn. **59**, 1252 (1990)
29. M.T. Dupuis, M. Viltange, Mikrochim. Acta 232 (1963)
30. I. Lindqvist, Acta Chem. Scand. **4**, 1066 (1950)
31. P. Caillet, P. Saumagne, J. Mol. Struct. **4**, 191 (1969)
32. S.A. Magarill, R.F. Klevtsova, Sov. Phys.-Crystallogr. **16**, 645 (1972)
33. B.M. Gatehouse, J. Less-Common Met. **36**, 53 (1974)
34. R.G. Bhattacharyya, S. Biswas, Inorg. Chim. Acta **181**, 213 (1991)
35. K. Stadnicka, J. Haber, R. Kozlowski, Acta Crystallogr. B **33**, 3859 (1977)
36. H.J. Becher, H.J. Brockmeyer, U. Prigge, J. Chem. Res. (S) 117 (1978); (M) 1670 (1978)
37. H.J. Becher, J. Chem. Res. (S) 92 (1980); (M) 1053 (1980)
38. B.M. Gatehouse, P. Leverett, J. Chem. Soc. Dalton 1316 (1976)
39. D.L. Sidebottom, Phys. Rev. B **61**, 14507 (2000)
40. J. Swenson, L. Borjesson, Phys. Rev. Lett. **77**, 3569 (1996)
41. H. Rawson, Phys. Chem. Glass. **1**, 170 (1960)
42. I. Stanworth, J. Soc. Glass Technol. **36**, 217 (1952)
43. Y. Dimitriev, St. Yordanov, L. Lakov, J. Non-Cryst. Solids **293–295**, 410 (2001)
44. M. Venkateswarlu, K. Narasimha Reddy, B. Rambabu, N. Satyanarayana, Solid State Ion. **127**, 177 (2000)
45. M. Venkateswarlu, N. Satyanarayana, B. Rambabu, J. Power Sources **85**, 224 (2000)
46. H.S. Chen, C.E. Miller, Rev. Sci. Inst. **41**, 1237 (1970)
47. K. Takada, J. Power Sources **394**, 74 (2018)
48. H. Pan, S. Zhang, J. Chen, M. Gao, Y. Liu, T. Zhu, Y. Jiang, Mol. Syst. Des. Eng. **3**, 748 (2018)
49. *Glass Nanocomposites; Synthesis, Properties and Applications* (Elsevier, 2016)
50. F. Zheng, M. Kotobuki, S. Song, M.O. Lai, L. Lu, J. Power Sources **389**, 198 (2018)
51. X. Yao, B. Huang, J. Yin, G. Peng, Z. Huang, C. Gao, D. Liu, X. Xu, Chin. Phys. B **25**, 018802 (2016)
52. E. Zanotto, Am. Ceram. Soc. Bull. **89**, 19 (2010)
53. S. Ojha, M. Roy, A. Chamuah, K. Bhattacharya, S. Bhattacharya, Mater. Lett. **258**, 126792 (2020)

Chapter 5
Experimental Tools for Characterizations of Lithium-Ion Doped Glassy Systems

Sanjib Bhattacharya

Abstract Various methods for characterization of as-prepared glassy samples have been discussed, and different features of their application are mentioned. Density-molar volume, FT-IR, FE-SEM, TEM, DSC, UV visible, and Raman spectroscopy are the different methods of material structural characterization. Among of them, infrared spectroscopy and Raman scattering are two important spectroscopic methods applied in the structural investigation of the local-order characterizing vitreous materials like oxide glasses. Electrical conductivity and modulus spectra are the key features of dielectric relaxation process.

Keywords XRD · FT-IR · UV–Vis · Raman spectroscopy · Electrical and dielectric properties

5.1 Introduction

Oxide glassy materials containing lithium are a part of non-crystalline solid materials which show short-range order, and some crystallinity may appear in a small volume [1]. While there has been a great amount of experimental work that has taken place in the area of glass-nanocomposites, a consensus has not yet been reached on how nanosized inclusions affect mechanical properties [2, 3]. Several attempts have been done to improve their mechanical and electrical properties [4–6]. Several techniques, both microscopic and macroscopic, have been developed for the study of the structure and properties of such glassy systems [2–6]. By measuring the viscosity, density, and electrical conductivity of such glass system, one can get an insight into their structures [7, 8]. Structural studies have been carried out by several investigators [9], using electron spin resonance (ESR), nuclear magnetic resonance (NMR), Raman, IR and Mossbauer spectroscopy, and X-ray diffraction.

Density-molar volume, FT-IR, FE-SEM, TEM, DSC, UV visible, and Raman spectroscopy are the different methods of material structural characterization. Among

S. Bhattacharya (✉)
UGC-HRDC (Physics), University of North Bengal, Darjeeling, West Bengal 734013, India
e-mail: ddirhrdc@nbu.ac.in; sanjib_ssp@yahoo.co.in

S. Bhattacharya and K. Bhattacharya (eds.), *Lithium Ion Glassy Electrolytes*,
https://doi.org/10.1007/978-981-19-3269-4_5

of them, infrared spectroscopy and Raman scattering are two important spectroscopic methods applied in the structural investigation of the local-order characterizing vitreous materials like oxide glasses [10, 11]. Oxide is one of the most common glass formers and is present in almost all commercially important glasses. It is often used as a dielectric material, and nanocomposites glasses possess scientific interest due to their technological interest [10, 11]. In nanocomposites glasses, CuI, AgI, CdI_2, MoO_3 are the basic glass former because of their higher bond strength, lower cation size, and smaller heat of fusion, so the structural investigation of these glasses is one of the most attractive points of glass formation and related doped systems. LiI, Li_2O, CuO, Se_2O, Ag_2O, MoO_3, V_2O_5, and ZnO can enter the glass network both as a network former and also as a network modifier, and due to this, the structure of this glass is expected to be different from that of phosphate and silicate glasses. The relationship between the microstructure, mechanical properties, and electrical properties of such glassy systems is needed to develop a better fundamental understanding of the behavior and properties of them, which is essential to contribute to an area of significant current technological interest [12, 13].

5.2 Methods Used for Characterization and Features of Their Application for Glass Composite Characterization

5.2.1 X-Ray Diffraction (XRD)

To ensure the nature of the prepared samples, X-ray diffraction is carried out on the powdered glassy samples using a Rich-Seifert X-ray diffractometer (model 3000P) for recording the diffraction traces (2θ versus intensity) of the powdered samples (Fig. 5.1). In this instrument, Ni-filtered CuK_α radiation operating at 35 kV and 25 mA in a step scan mode was used. The step size was taken to be of 0.02° in 2θ and a hold time of 2 s per step. The hardware of XRD 3000 systems comprises of the generator ID 3000, the monitor, accessory controller C 3000, and the timer/counter hardware. The diffraction traces were recorded at room temperature. From the diffraction peaks of the XRD pattern, the average particle size of different nanoparticles was determined using Scherer formula [14]

$$t = 0.89\lambda/(\beta\cos\theta) \tag{5.1}$$

where t denotes the average grain size of the particles, λ stands for the X-ray wavelength (1.54 Å), θ for the Bragg's diffraction angle, and β is for the peak width in radians at half-height.

Crystallite size and the lattice strain of the crystal can be evaluated separately from XRD study by Hall's equation [15],

Fig. 5.1 Experimental set-up for Rich-Seifert X-ray diffraction measurements

$$\beta_{\mathrm{hkl}} \cos\theta/\lambda = 2\eta \sin\theta/\lambda + K/D \tag{5.2}$$

where β_{hkl} is the full width half maximum of a given (hkl) diffraction peak, λ is the wavelength of the X-ray, D is the crystallite size, η is the measure of the heterogeneous lattice strain, θ is the Bragg's angle, and K is a constant of 0.9. From the plots between $\beta_{\mathrm{hkl}}\mathrm{Cos}\theta/\lambda$ and $\mathrm{Sin}\theta/\lambda$ for the major (hkl) diffraction peaks of different nanocrystals in the samples, the values of β_{hkl} were determined from the Gaussian function which was fitted to the major diffraction peaks. The values of η, the measure of lattice strain, were obtained from the slope of the plots.

When the sample and the detector rotate, the strength of the reflected or diffracted X-rays is documented. When the geometry of the X-ray event is affected, the sample satisfies the Bragg's equation, and constructive or destructive interference and peak intensity occur. The detector detects and develops this X-ray signal and translates the signal to read speed, which is then connected to output device [14, 15]. The X-ray diffractometer geometry is such that the sample moves along the X-ray-accumulated beam path at an angle, while the X-ray detector is attached to the armrest to accumulate diffraction X-rays and rotates at an angle of 2θ [14, 15]. The tool for supporting the angle and rotation of the sample is known as goniometer. For characteristic powder diffraction patterns, data at 2θ from angles ~5° to 80° are embedded in X-ray scanning [14, 15].

The identification of crystal structure of unknown materials has been performed usually with a powder XRD [14, 15]. Powder sample is made with fine grains with a single crystal/polycrystal structure, but oriented randomly, when exposed to X-rays, and hence, the diffraction intensity is assumed the sum of the X-rays reflected from all the fine grains. A standard database is maintained by the International Centre for Diffraction Data (ICDD) and Joint Committee on Powder Diffraction Standards (JCPDS) for the identification of inorganic and organic materials with a crystal structure. XRD patterns can be broadly employed for (i) identification of crystalline phases and their degree of crystallization, (ii) estimating crystallite size, and (iii) residual strain [14, 15]. A single crystal specimen in a XRD diffractometer would produce only one family of sharp peaks in the diffractogram. Amorphous glassy materials generally exhibit a broad "halo" near 20°, indicating no long-range order and are supposed to be macroscopically isotropic. In a mixture of amorphous and crystalline material (polycrystalline), the XRD pattern may exhibit both sharp and broad features, where the sharp peaks are due to the crystalline component/phase and the broad features are owing to the amorphous phase. Deconvolution of the mixture of XRD spectrum into separate XRD spectrum with sharp-only diffraction peaks and broad-only diffraction peaks may be taken into consideration for estimating the percentage of amorphous content in the compositions [15]. It is also noted by observing various diffraction peaks of different samples that the diffraction peak shape is found to be affected due to the size of crystallites [14, 15]. As the crystallite size is reduced, the diffraction peaks broaden to an extent to merge into each other to form a single broad diffraction peak.

5.2.2 *Field Emission Scanning Electron Microscopy (FE-SEM) and Energy-Dispersive X-Ray Spectroscopy (EDS)*

To explore the microstructure and surface morphology of the prepared glasses and glass-nanocomposites, field emission scanning electron micrographs (FE-SEM) of the polished surfaces of the samples were taken in a field emission scanning electron microscope (JEOL JSM-6700F) [16, 17]. A thin platinum coating (~150 Å) was deposited on the polished surfaces of the samples by vacuum evaporation technique for a conducting layer. The quantitative investigation of the final compositions has been done from EDS study of the corresponding FE-SEM image (Fig. 5.2).

A tungsten filament is heated to generate the electron beam. FE-SEM beam of particles is generated by the incident electron in a column above the specimen chamber. Incident electrons' energy can be as low as 100 eV or as high as 30 keV depending upon the purpose of the study. Electrons are concentrated in a small beam of electromagnetic lenses in the FE-SEM column beam. The scanning coils are placed on the edges of the column, and beam is centered directly on the sample of the surface. The electron beam is concentrated in one place and scanned along a line for X-ray analysis.

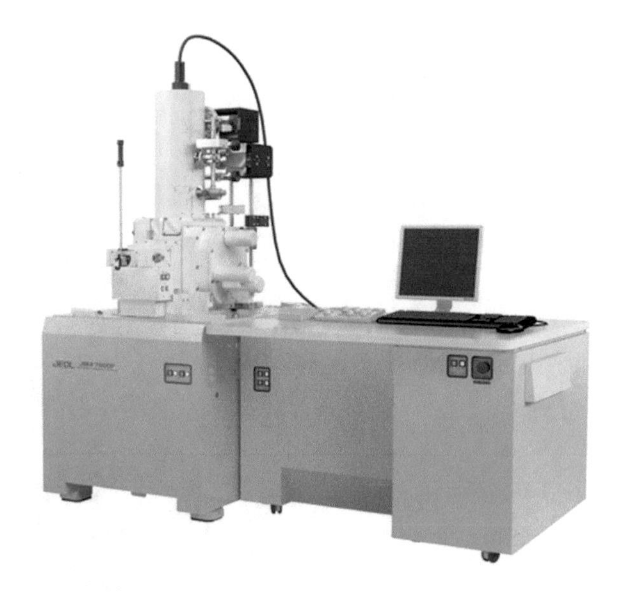

Fig. 5.2 Experimental set-up for field emission scanning electron microscope (JEOL JSM-6700F)

5.2.3 Transmission Electron Microscopy (TEM)

To explore the microstructure of the prepared glasses, transmission electron micrographs of the samples prepared in the form of thin films were taken in a transmission electron microscopic (JEOL JEM2010). The samples were first powdered in a mortar, then sonicated for 20 min in acetone to form very tiny particle. Thus, a very dilute colloidal solution in acetone is formed. Then, pouring a very small drop of that colloidal solution on carbon-coated grid (300 meshes), they were made suitable for transmission electron microscopy. The grids thus prepared were used to obtain the microstructure of the samples. Microstructure thus obtained shows nanocrystalline nature in some of the samples (Fig. 5.3).

The electron beam in the electronic gun targets a small, thin, and coherent beam through a condenser lens [18, 19]. The beam at that time hits the sample on the grid, and parts of it are transmitted depending on the thickness and electron transparency of the sample [18, 19]. The objective lens focuses this ported part on a camera image with a phosphor screen or charger (CCD) [18–20]. Possible objective openings can be used to enhance contrast by blocking high angle diffraction electrons [21]. It then passes through the column using the intermediate projection lenses to enlarge the image [21]. Dark areas of the image are the sections of the sample in which fewer electrons are transmitted, and the lighter parts of the image areas are characterized by a sample of more electrons transmitted through [18, 19, 21]. Another aspect of TEM is to take selected area electron diffraction pattern (SAED) micrograph, which can be used to find out inter-planner spacing of the crystalline planes. Specimens under study with ~100 nm thick can allow the electrons of 100–400 keV through them easily.

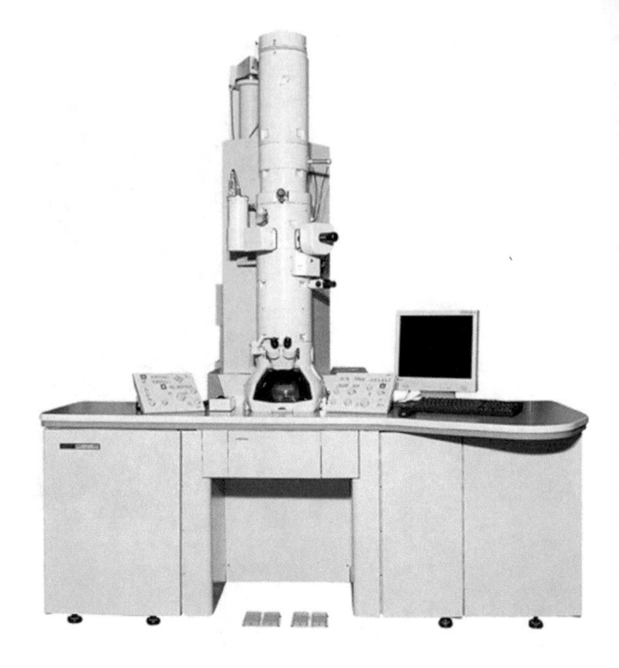

Fig. 5.3 Experimental set-up for transmission electron microscopy (JEOL JEM2010)

5.2.4 *Differential Scanning Calorimetry (DSC)*

To determine the thermodynamic properties such as glass transition temperature, crystallization temperature, and melting temperature, differential scanning calorimetry (DSC) of the powered glass samples is performed in a PerkinElmer. Differential scanning calorimeter (DSC 7) operates in the temperature range from − 150 °C to 500 °C in nitrogen atmosphere. In these processes, fine powders of sample is taken in aluminum crucible. Pure Al_2O_3 in other crucible is used as a reference. The difference between the glass transition temperature T_g and the crystallization temperature T_c, which is considered as a measure of thermal stability against crystallization, ranges from 35 °C to 40 °C for different compositions, and thus the glasses and glass-nanocomposites can be regarded as stable.

5.2.5 *Fourier Transform Infrared Spectroscopy (FT-IR)*

The FT-IR spectra of the powdered samples in KBr matrices in transmission mode were recorded in a Nicolate FT-IR spectrophotometer (Magna IR-750, Series II) in the wave number range of 400–4000 cm^{-1}at a temperature 25 °C and humidity at 50–60%. Each bonding in the sample has a characteristic frequency of vibration. When IR is transmitting through the sample, the frequency of the bonding of the

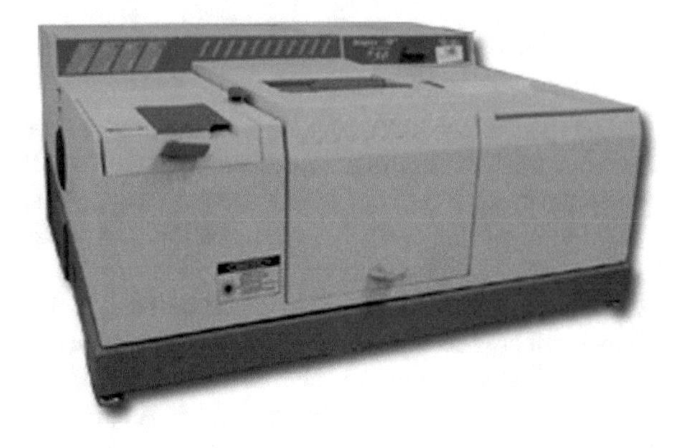

Fig. 5.4 Experimental set-up for Nicolate FT-IR spectrophotometer

sample is exactly matched with particular frequency of the IR region, then some peaks are found due to resonance of frequency (Fig. 5.4).

5.2.6 *Ultraviolet Visible Spectroscopy (UV–Vis)*

To explore the optical band gap of the prepared glasses and glass-nanocomposites, ultraviolet visible spectroscopy (UV–Vis) of the samples was taken in a spectrometer [22]. UV–Vis absorption studies were made in the reflection mode using PerkinElmer Lamda-750 spectrophotometer (Fig. 5.5). The colloidal suspensions using ethanol as solvent for the absorption studies were prepared with the help of an ultrasonic homogenizer (Takashi SK-500F).

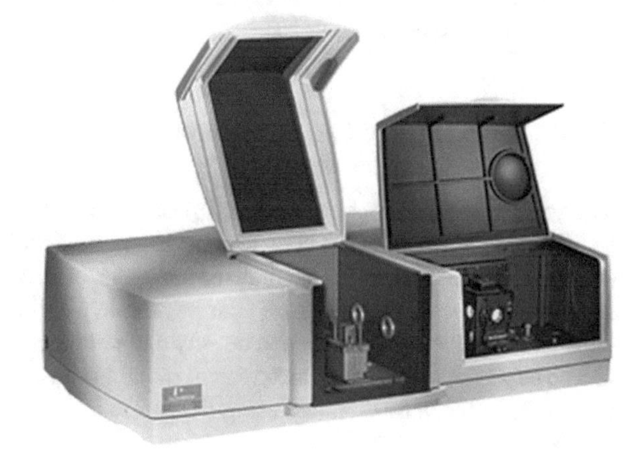

Fig. 5.5 Experimental set-up for UV–Vis. spectroscopy (PerkinElmer Lamda-750)

5.2.7 Raman Spectroscopy

Raman spectroscopy [20, 23] has great importance for exploring the molecular structure and composition of inorganic materials. The Raman scattering arises when monochromatic light irradiates a sample causing a small portion of the scattered radiation to exhibit shifted frequencies that are comparable to the sample's vibration transitions. In molecular systems, wavelengths are mainly in ranges associated with transitions between rotational, vibrational, and electronic levels [20, 23]. The Raman effect is established in molecular distortions in the electric field persistent to molecular polarization (α). The laser beam is an oscillating electromagnetic wave with an electric vector interacting with the sample, and it prompts the electric dipole moment $P = \alpha E$ that deforms molecules. As an outcome of the periodic deformation, molecules begin to vibrate at a distinctive frequency. Specifically, monochromatic laser light often excites the molecules and converts them into oscillating dipoles. These oscillating dipoles release light from three different frequencies:

(a) A molecule that does not use Raman active modes absorbs a photon frequently ν_0. The agitated molecule yields the same fundamental vibratory state and radiates light with the same frequency ν_0 as the source of excitation, which is known as elastic Rayleigh's scattering [20, 23].
(b) Photon with frequency ν_0 absorbs by the Raman active molecule, which is during the interaction in elemental vibrational state. Part of the photon energy is transmitted to Raman active mode, and consequently, frequency of scattered light decreases (Stokes lines) [20, 23].
(c) A photon frequency ν_0 absorbs by a Raman active molecule that is already in an agitated vibrational state during the interaction. The excess energy is released from the excited active Raman mode, the molecule yields the fundamental vibrational state, and consequently, frequency of the scattered light increases (anti-Stokes lines) [20, 23].

5.2.8 Density and Molar Volume

The densities of the prepared glass and glass-nanocomposite samples were measured by Archimedes's principle using acetone as an immersion liquid. Molar volume of a substance is the volume of one mole of that substance. Molar volume of a substance is defined as the ratio of its molecular or atomic weight whichever is suitable to its density. The relationship between density and composition of an oxide glass system can be expressed in terms of an apparent molar volume of oxygen (V_{M}) for the glass system, which can be obtained using the formula

$$V_{\mathrm{M}} = \sum x_i M_i / \rho \tag{5.3}$$

where x_i is the molar fraction and M_i is the molecular weight of the ith component.

Fig. 5.6 Some polished glass-nanocomposite samples for hardness testing

5.2.9 Microhardness Testing

There are two types of microhardness testing process, namely Knoop and Vickers hardness test which have a very small diamond indenter of pyramidal geometry and forced into the surface of the specimen. Measuring methods and machines have been appropriately adapted to perform hardness measurements on very small objects (glass-nanocomposites), very thin ~1 mm thick [24, 25]. Micro-indentation hardness, also often called microhardness, means the determination of hardness values with low test forces. Compared to the macrohardness testing methods described previously, the test force range in microhardness testing is very small. Following the ASTM Standard Test Method for Micro-indentation Hardness of Materials (E 384), the range was between 1 and 200 gf (9.8×10^{-3} and 9.8 N), and the indentations were correspondingly very small. The hardness value that has already been described is in any case ascertained by dividing the test force by the remaining indentation surface area (Fig. 5.6).

5.2.10 Electrical and Dielectric Property: Measurement Techniques

The prepared glass and glass-nanocomposites were shaped rectangular or circular by cutting the samples with a diamond cutter. Silver paste was deposited on both surfaces of the polished samples as electrodes. The silver-pasted samples were then heated at 50–60 °C for two hours for the stabilization of the electrodes. Electrical properties of materials are determined through four fundamental parameters called dielectric constant, tangent of dielectric loss angle, dielectric breakdown, and electrical conductivity [26].

The block diagram for the experimental set-up for electrical measurements is shown in Fig. 5.7. The sample is kept in the sample holder into the furnace (*F*).

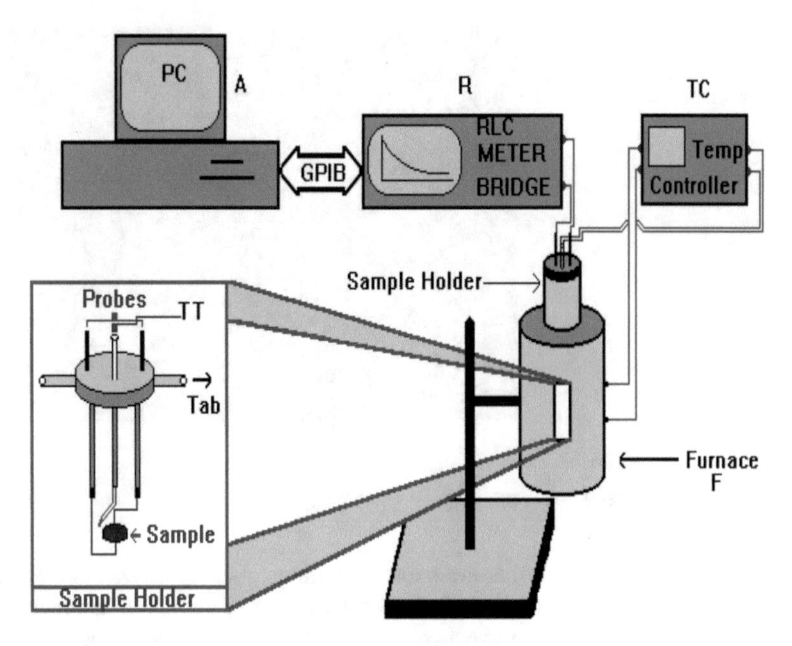

Fig. 5.7 Block diagram for the experimental set-up for electrical measurements

The sample is connected to a LCR Meter Bridge (*R*) (Hioki made) through the two probes (as seen in the Fig. 5.8) coming out of the sample holder. A thermocouple TT is inserted in the chamber and kept near to the sample. The other end of the thermocouple is connected to the temperature controller (TC). Here, a personal computer (PC) is interfaced with the LCR Meter Bridge through GPIB card. So, the LCR Meter Bridge can be run with the instructions by the PC. Measurements such as capacitance and conductance can be carried out as a function of frequency of the sample at different temperatures. The AC measurements can be made in the frequency range 42 Hz–5 MHz.

The frequency-dependent AC conductivity $\sigma^{/}(\omega)$ at frequency ω is determined from the following relation

$$\sigma^{/}(\omega) = (t/A)G(\omega) \tag{5.4}$$

The real part of the permittivity $\varepsilon^{/}(\omega)$ can be related to the capacitance by

$$\varepsilon^{/}(\omega) = (C(\omega)/\varepsilon_0) \times (t/A) \tag{5.5}$$

and the imaginary part $\varepsilon^{//}(\omega)$ of the permittivity was related to the real part of the conductivity

$$\varepsilon^{//}(\omega) = tG(\omega)/A\omega\varepsilon_0 \tag{5.6}$$

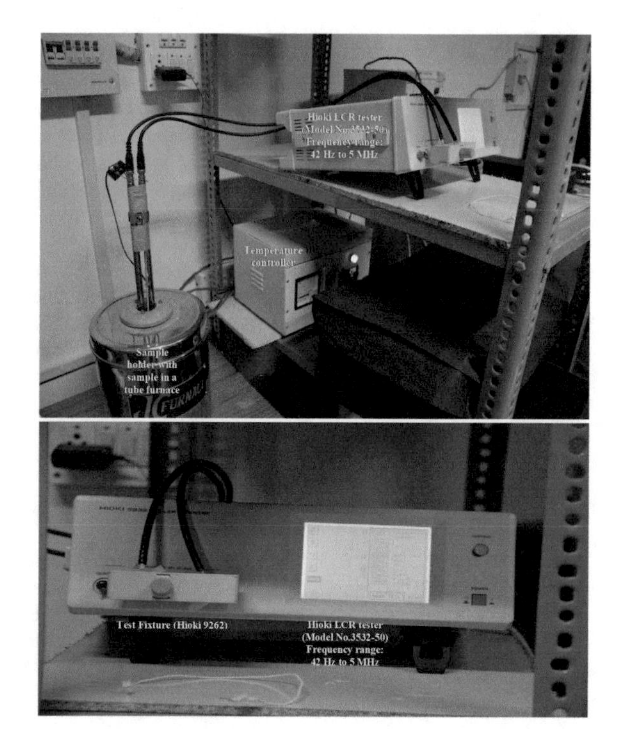

Fig. 5.8 Experimental set-up for electrical measurements

where ε_0 is the free space permittivity having value of 8.8514×10^{-12} F/m. The real and imaginary parts of the electric modulus $[M^*(\omega) = M^{/}(\omega) + iM^{//}(\omega)]$ are further calculated from the real and imaginary parts of the permittivity by the following relation

$$\mathrm{M}^{/}(\omega) = \varepsilon^{/}(\omega) / \left\{ \left|\varepsilon^{/}(\omega)\right|^2 + \left|\varepsilon^{//}(\omega)\right|^2 \right\} \tag{5.7}$$

$$\mathrm{M}^{//}(\omega) = \varepsilon^{//}(\omega) / \left\{ \left|\varepsilon^{/}(\omega)\right|^2 + \left|\varepsilon^{//}(\omega)\right|^2 \right\} \tag{5.8}$$

References

1. S.C. Baidoc, Ph.D. Thesis summery, “Babeş-Bolyai” University (2011)
2. D. Gersappe, Phys. Rev. Lett. **89**(5), 058301 (2002)
3. J. Jordan et al., Experimental trends in polymer nanocomposites—a review. Georgia Institute of Technology, pp. 1–29
4. P.H.T. Vollenberg, D. Heikens, Polymer **30**, 1656 (1989)
5. C.M. Chan, J. Wu, J.X. Li, Y.K. Cheung, Polymer **43**, 2981 (2002)
6. C.L. Wu, M.Q. Zhang, M.Z. Rong, K. Friedrich, Compos. Sci. Technol. **62**, 1327 (2002)
7. S. Barth, A. Feltz, Solid State Ion. **34**, 41 (1989)

8. M.D. Ingram, Phys. Chem. Glasses **28**, 215 (1987)
9. B.N. Meera, A.K. Sood, N. Chandrabhas, J. Ramakrishna, J. Non-Cryst. Solids **126**, 224 (1990)
10. A.K. Hassan, L.M. Torell, L. Börjesson, H. Doweidar, Phys. Rev. B **45**, 12797 (1992)
11. Y.D. Yiannopoulos, G.D. Chryssikos, E.I. Kamitsos, Phys. Chem. Glasses **42**, 164 (2001)
12. A.A. Alemi, H. Sedghi, A.R. Mirmohseni, V. Golsanamlu, Bull. Mater. Sci. **29**, 55 (2006)
13. Q. Zuqiang, Ph. D. Thesis, B. S. University of Science and Technology Beijing, China (2004)
14. H.M. Xiong, X. Zhao, J.S. Chen, J. Phys. Chem. B **105**, 10169 (2001)
15. W.H. Hall, J. Instrum. Methods **75**, 1127 (1950)
16. N.L. Donovan, G.H. Chandler, S. Supapan, *Scanning Electron Microscopy* (Wiley, 2002)
17. A. Bogner, P.H. Jouneau, G. Thollet, D. Basset, C. Gauthier, Micron **38**, 390 (2007)
18. C. Carter Barry, B.W. David, *Transmission Electron Microscopy: A Textbook for Materials Science* (Springer, 2012), pp. 3–89. ISBN: 978-0-387-76500-6
19. C. Carter Barry, B.W. David, *Transmission Electron Microscopy Diffraction, Imaging, and Spectrometry* (Springer, 2011), pp. 1–166
20. E. Smith, G. Dent, Modern Raman spectroscopy—a practical approach. ISBN 0-471-49668-5
21. G. Thomas, M.J. Goringe, *Transmission Electron Microscopy of Metals* (Wiley, New York, 1979)
22. R. Bhargava, Infrared spectroscopic imaging: the next generation. Appl. Spectrosc. **66**, 1091 (2012)
23. J.R. Ferraro, K. Nakamoto, C.W. Brown, *Introductory Raman Spectroscopy*, 2nd edn. ISBN: 978-0-12-254105-6
24. J. Pelleg, *Mechanical Properties of Materials 190* (Springer, Dordrecht, 2013)
25. I. Brooks, P. Lin, G. Palumbo, G.D. Hibbard, U. Erb, Mater. Sci. Eng. A **491**, 412 (2008)
26. A.R. Hippel, *Dielectrics and Waves* (Willey, New York, 1954)

Part II
Features of Some Lithium Doped Glassy Systems and Properties

Chapter 6
DC Electrical Conductivity as Major Electrical Characterization Tool

Amartya Acharya, Koyel Bhattacharya, Chandan Kr Ghosh, and Sanjib Bhattacharya

Abstract Various transport theories regarding transport process in glassy matrices have been discussed. Li_2O-doped glassy ceramics have been prepared using melt-quenching route, and their electrical DC conductivity has been studied in wide temperature regime. It is anticipated from the nature of composition that Li^+ conduction mostly contributes to electrical conductivity at high temperature, and Mott's variable-range hopping (VRH) model has been utilized to analyze low-temperature DC conductivity data due to polaron hopping. Composition-dependent DC conductivity is also discussed.

Keyword DC conductivity · Mott's variable-range hopping (VRH) model · Anderson-Stuart model

6.1 General Consideration

Ionic transport processes in glasses and glass-nanocomposites have been a subject of deep scientific interest [1–3] for more than half a century, and to date, a large number of glass-forming systems had been investigated over a wide range of compositions with respect to the transport-related properties such as conductivity and its dependence on temperature and frequency, radio tracer diffusion, NMR, electrical and mechanical relaxation, etc.

A. Acharya · S. Bhattacharya (✉)
UGC-HRDC (Physics), University of North Bengal, Darjeeling, West Bengal 734013, India
e-mail: ddirhrdc@nbu.ac.in; sanjib_ssp@yahoo.co.in

Composite Materials Research Laboratory, UGC-HRDC (Physics), University of North Bengal, Darjeeling, West Bengal 734013, India

K. Bhattacharya
Department of Physics, Kalipada Ghosh Tarai Mahavidyalaya, Bagdogra, Darjeeling, West Bengal 734014, India

A. Acharya · C. K. Ghosh
Department of Electronics and Communication Engineering, Dr. B. C. Roy Engineering College, Durgapur, West Bengal 713026, India

S. Bhattacharya and K. Bhattacharya (eds.), *Lithium Ion Glassy Electrolytes*,
https://doi.org/10.1007/978-981-19-3269-4_6

The glasses possess certain advantage over their crystalline counterparts, which includes physical isotropy, the absence of grain boundaries, good workability and continuously variable composition. Moreover, the room temperature conductivity in these glasses can vary from as little as 10^{-15} to as much as $10^{-2}\Omega^{-1}$ cm^{-1}, which makes them suitable for many electrochemical applications. The latter values of conductivity arise in fast-ion conductors in which the diffusing atoms are charged and carry electric current. This ionic contribution to the electric current exceeds the contribution from electrons. Various types of ions can diffuse in glasses which include the Li^+ ion (the smallest) to Ag^+ ion (the most deformable) having the highest conductivity [2, 3]. Some excellent review on the dynamic properties of the ions in the glasses can be found in the works of Ingram, Angell and Kahnt [3–5].

6.2 Transport Theory with Examples

The discovery of fast-ion conduction in oxide glass [6] and the energy crisis of the early 1970s stimulated much interest in using glasses as solid-state electrolytes in advanced battery systems. The microscopic mechanisms responsible for ionic conduction in glasses, however, are still not well understood due to the difficulty in independently determining the carrier concentration and mobility. The DC and AC conductivities of the ionically conducting glasses have been studied extensively for traditional glass formers [7–9]. The DC conductivity for materials with one type of carrier is given by

$$\sigma_{\mathrm{dc}} = (\mathrm{Ze})n\,\mu \tag{6.1}$$

where Ze is the charge of the carrier, n is the concentration of mobile carriers, and μ is the mobility. The concentration of mobile ions may be thermally activated and can be written as

$$\begin{aligned} n &= N_0 \exp(-\Delta G_c/k_\mathrm{B}T) \\ &= N_0 \exp(-\Delta S_c/k_\mathrm{B}) \exp(-\Delta H_c/k_\mathrm{B}T) \\ &= N_e \exp(-\Delta H_c/k_\mathrm{B}T) \end{aligned} \tag{6.2}$$

where ΔG_c is the free energy necessary to impart a carrier population, ΔS_c is the associated entropy, ΔH_c is the enthalpy, k_B is the Boltzmann constant, T is the absolute temperature, and N_e is the effective infinite temperature ion concentration which includes the entropy term. The mobility is related to the diffusivity (D) through Nernst-Einstein relation

$$\begin{aligned} \mu = \mathrm{Ze}D/k_\mathrm{B}T &= \mathrm{Ze}\gamma\lambda^2\upsilon_\mathrm{H}/k_\mathrm{B}T \\ &= \left(\mathrm{Ze}\gamma\lambda^2\upsilon_0/k_\mathrm{B}T\right)\exp(-\Delta G_m/k_\mathrm{B}T) \end{aligned}$$

$$= \left(Ze\gamma\lambda^2 v_0/k_B T\right)\exp(-\Delta S_m/k_B)\exp(-\Delta H_m/k_B T)$$
$$= \left(Ze\gamma\lambda^2 v_e/k_B T\right)\exp(-\Delta H_m/k_B T) \tag{6.3}$$

where ΔG_m is the free energy for ion migration, ΔS_m is the associated entropy, ΔH_m is the enthalpy, γ is the geometrical factor for ion hopping, λ is the average hop distance between the mobile ion sites, ν_H is the hopping frequency, ν_0 is the jump attempt frequency of the ion, and ν_e is the effective jump attempt frequency including the entropy term. From the first law of thermodynamics under conditions of constant specimen volume and temperature, the enthalpy and energy state functions are equal. Thus, we replace ΔH by ΔE. Therefore, substituting Eqs. (6.2) and (6.3) to (6.1) yields

$$\sigma_{dc} = \left(N_e(Ze)^2\gamma\lambda^2 v_e/k_B T\right)\exp\{-(\Delta E_c + \Delta E_m)/k_B T\} \tag{6.4}$$

which agrees quite well with the experimental results in the limited temperature range.

Any discussion regarding the mechanism of ion transport in glasses must focus on two themes.

- Strong electrolyte theories (Anderson-Stuart model)
- Weak electrolyte theories (Ravaine-Souquet model).

6.2.1 Anderson-Stuart Model

The Anderson-Stuart [9] model is a structural model which considers the activation energy as the energy required to overcome electrostatic forces (ΔE_B) plus the energy required to open up "doorways" in the structure large enough for the ion to pass through (ΔE_S). An atomic-level representation of this model by Martin and Angell [8] is shown in Fig. 6.1.

According to Anderson-Stuart model, the activation energy is

$$\Delta E_{act} = \Delta E_B + \Delta E_S \tag{6.5}$$

where the binding energy term is given as

$$\Delta E_B = \frac{ZZ_0 e^2}{\varepsilon_\infty}\left[\frac{1}{r + r_0} - \frac{2}{\lambda}\right] \tag{6.6}$$

and the strain energy term is given as

$$\Delta E_s = \pi G_D \lambda (r - r_D)^2/2 \tag{6.7}$$

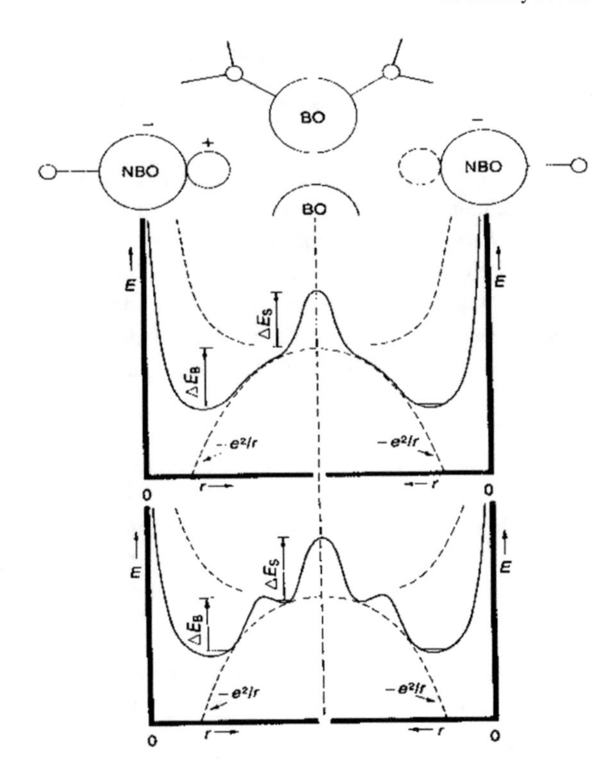

Fig. 6.1 Representation of the energies of cation conduction process according to the Anderson-Stuart model as interpreted by Martin and Angell [8]. Republished with permission from Martin and Angell [8]

In these equations, G_D is the shear modulus of the glass, and r_D, r and r_0 are the interstitial window, the mobile cation and the non-bridging anion radii, respectively; λ is the average jump distance; Z and Z_0 are the number of charges on the mobile cation and the anion; and e is the charge on the electron [10, 11].

To make the Anderson-Stuart model more realistic, minor changes to the strain term have been proposed by McElfresh and Howitt [10]. Elliott [11] has pointed out that the Anderson-Stuart model neglects specific polarization and repulsion terms and includes these terms in the overall Coulomb potential. Other models describing the activation energy have been suggested, but nearly all follow the general principles of the Anderson-Stuart model.

The physical and structural parameters necessary to verify the validity of the Anderson-Stuart model and experimental methods for measuring these parameters are:

- ΔE_{act}: determined from wide temperature-range ionic conductivity measurements
- λ: approximated from NMR static linewidth measurements or density measurements
- G: determined from acoustic measurements
- r_D: the interstitial window radius determined from inert gas diffusion studies.

To test the validity of the Nernst-Einstein derivation for pre-exponent term, the following additional parameters are needed:

- σ_0: determined from wide temperature-range ionic conductivity measurements
- ν_0: determined from the Far-IR ion vibrational frequency
- γ: usually taken as approximately equal to $1/6$.

6.2.2 *Ravaine-Souquet Model*

The correlation between ionic conductivity and thermodynamic activity is the basis of weak electrolyte or Ravaine-Souquet model [12]. In glasses, the addition of M_2O or M_2S typically results in the added anions, becoming part of the glass structure by covalently bonding to the glass-forming cations, while the added alkali cations reside in a local region supporting charge neutrality. Most of these alkali cations are unionized and immobile, but a small fraction may dissociate from these sites to form ionized or dissociated "mobile" cations. These ions are proposed to contribute to the ionic conduction. The formation of mobile cations M^+ from associated oxide complex in a glass is taken analogous to the dissociation of modifier salt added to the glass [13].

$$M_2O = M^+ + OM^- \tag{6.8}$$

and the concentration independent dissociation constant is given by

$$K = [M^+][OM^-]/[M_2O] \tag{6.9}$$

From Eq. (6.8), the $[M^+]$ and $[OM^-]$ are equal, and the concentration of the free dissociated ion is, therefore, given by

$$\begin{aligned} [M^+] &= K^{1/2}[M_2O]^{1/2} \\ &= K^{1/2}\left[a_{M_2O}\right]^{1/2} \end{aligned} \tag{6.10}$$

where a_{M2O} is the thermodynamic activity equated to the M_2O concentration. The equality is valid for very dilute solutions (Henry's law). The ionic conductivity is proportional to the concentration of mobile ions, so that

$$\sigma \propto [M^+] = K^{1/2}\left[a_{M_2O}\right]^{1/2} \tag{6.11}$$

Ravaine and Souquet performed both ionic conductivity and concentration cell emf measurements for various sodium and potassium silicate glasses. They plotted conductivity ratio versus activity ratio for various pairs of glasses and showed that the slope was ½ for this log plot.

6.3 DC Electrical Conductivity of Some Li Containing Glassy Systems Using Various Models

Glassy ceramics $x\mathrm{Li_2O}$–$(1-x)$ $(0.8\mathrm{V_2O_5}$–$0.2\mathrm{ZnO})$ with $x = 0.1$, 0.2 and 0.3 have been developed in the laboratory by solid-state reaction. Complex impedance plots for $x = 0.1$ are presented in Fig. 6.2a at various temperatures. The DC electrical conductivity (σ_{dc}) has been computed from the semicircular portions of Fig. 6.2a. It is noted in Fig. 6.2a that grain boundary resistance as well as polarization effects are absent. Similar results are obtained for other samples. Yuan et al. [14] reveal that a small amount of doping is sensitive to the purity of the sample, which is reflected in the nature of the plot. Generally, the AC response of the system indicates a relation between the applied voltage and the current through the sample under consideration. The equivalent circuit containing ideal resistive and reactive components is presented in the inset of Fig. 6.2a. It may be proposed to explore AC response of the system.

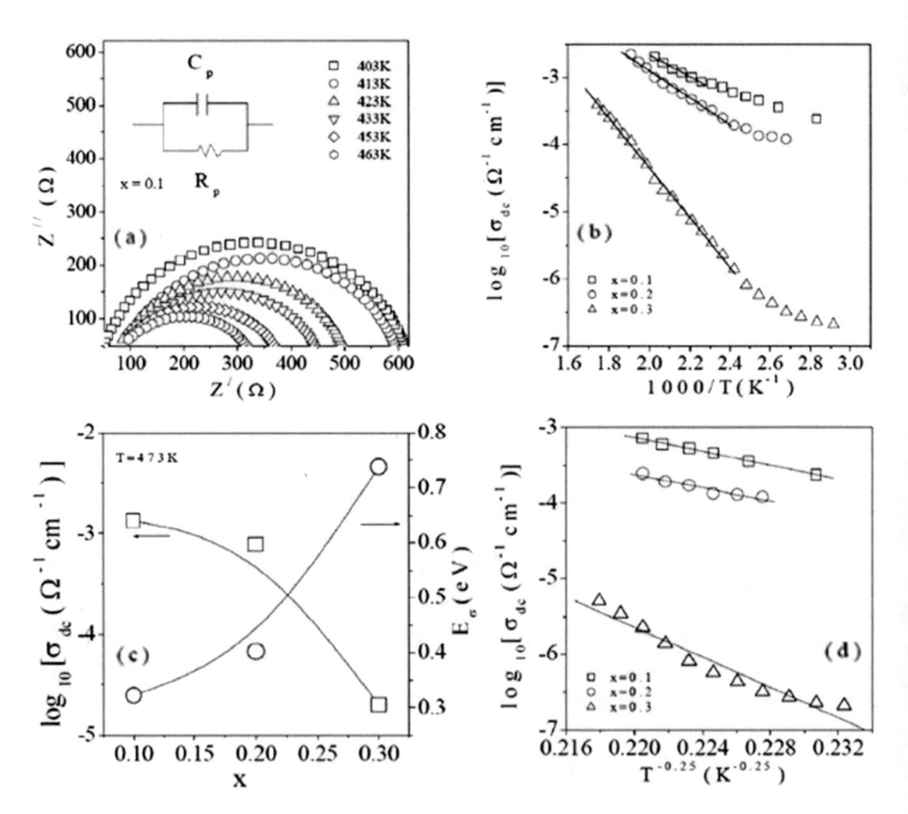

Fig. 6.2 **a** Cole–Cole plot of resistivity and corresponding equivalent circuit for measurement **b** temperature dependency of DC conductivity; **c** fixed-temperature (473 K) DC conductivity and activation energy and **d** low-temperature DC conductivity plot using Mott's model. Republished with permission from Acharya et al. [15]

In the equivalent circuit (parallel RC), the overall AC impedance of the present circuit can be represented as [14]:

$$\begin{aligned}\frac{1}{z} &= \frac{1}{Z_R} + \frac{1}{Z_C} \\ &= \left(\frac{1}{R} + j\omega C\right)^{-1} \\ &= \frac{R}{1 + j\omega RC} \\ &= \frac{R}{1 + (\omega RC)^2} - j\frac{\omega R^2 C}{1 + (\omega RC)^2}\end{aligned} \tag{6.12}$$

where Z_R and Z_C are the resistive and reactive components.

This result directly indicates the form of real (Z_{re}) and imaginary (Z_{im}) impedances of the parallel RC circuit as:

$$\begin{aligned}Z_{re} &= \frac{R}{1 + (\omega RC)^2} \\ Z_{im} &= -\frac{\omega R^2 C}{1 + (\omega RC)^2}\end{aligned} \tag{6.13}$$

and the phase angle φ can be presented as:

$$\tan\varphi = -\omega RC \tag{6.14}$$

At low frequency ($\omega RC \ll 1$), $Z_{re} \approx R$ and $Z_{im} \approx 0$. This result implies that this RC circuit acts as a resistor. On the other hand, at high frequency ($\omega\text{RC} \gg 1$), $Z_{re} \approx 0$ and $Z_{im} \approx 1/\omega C$, and the present circuit acts as a capacitor with time constant, equal to RC.

Equations (6.12) and (6.13) yield

$$\left(Z_{re} - \frac{R}{2}\right)^2 + Z_{im}^2 = \left(\frac{R}{2}\right)^2 \tag{6.15}$$

Equation (6.15) indicates a half-circle in the complex plane, with a radius of *R*/2, which can be validated by Fig. 6.2a.

Figure 6.2b shows the variation of DC conductivity with reciprocal temperatures, which demonstrates thermally activated nature. It may be anticipated from the nature of variation of DC conductivity with temperature in Fig. 6.2b that the present glassy system must contain both ionic and electronic components, which may cause the total conductivity. However, semiconducting properties may arise due to presence of a small percentage of transition metals (vanadium), via polaron hopping, from lower cationic valence state to higher valence state [16, 17]. Similarly, ionic conductivity may arise due to transport of lithium ions in present glassy matrix, which may impart electrical conductivity of the present system [18].

6.4 Study of Temperature and Composition Dependency of Conductivity

It is also noteworthy from Fig. 6.2b that DC conductivity decreases with Li_2O content in the compositions. It is quite clear from Fig. 6.2b that low-temperature DC conductivity data may arise due to polaron hopping process [19] and high-temperature DC conductivity data have been received mainly due to conduction of Li^+ ions [20]. Here, the polaron conduction is achieved by the following conversion [21]:

$$V^{+5} \rightarrow V^{+6} + \text{electron}$$

Here, ZnO acts as a stabilizer [22]. As the Li_2O content increases in the compositions (V_2O_5 decreases), more and more number of V^+ ions are expected to take part into bonds of network, thereby contributing less number of polaron in the conduction process. As a consequence, conductivity drops down at low temperature.

High-temperature DC conductivity data as shown in Fig. 6.2b are found to increase with temperature linearly, which may be analyzed using Arrhenius equation:

$$\sigma_{\mathrm{dc}} = \sigma_0 \exp(-E_\sigma / kT) \tag{6.16}$$

where E_σ is the DC activation energy for present glassy ceramics under investigation, T is the absolute temperature, and k is the Boltzman constant. Figure 6.2c depicts the variation of σ_{dc} at 473 K with compositions. It is interestingly noted that the sample with $x = 0.1$ shows the highest DC conductivity. DC conductivity is found to decrease as the Li_2O content increases in the compositions. Computed activation energy corresponding to σ_{dc}, obtained from the slopes of the best-fitted straight lines of Fig. 6.2a, is also presented in Fig. 6.2c, which shows opposite nature of DC conductivity.

To interpret DC conductivity data in low temperature ranges (below half of the Debye temperature), Mott's variable-range hopping (VRH) model [23, 24] has been considered.

Here, a charge carrier (polaron) hops from one localized state to another. It is also assumed here that the density of states is finite and localized at the Fermi level. Mott's VRH conductivity [23, 24] may be expressed as:

$$\sigma_{\mathrm{dc}} = A \exp\left[-\left({T_0}/{T}\right)^{0.25}\right] \tag{6.17}$$

Table 6.1 Estimated values of density of states near Fermi level ($N(E_{FM})$) and fixed-temperature DC conductivity of xLi$_2$O–(1−x)(0.5V$_2$O$_5$–0.5ZnO) glass–ceramics. Estimated errors of measurements are mentioned

x	$N(E_{FM})$ (eV^{-1} cm^{-3}) (± 0.01)	log$_{10}$[σ_{dc}(Ω^{-1} cm^{-1})] at 357 K (± 0.01)	log$_{10}$[σ'(Ω^{-1} cm^{-1})] at 4.8 MHz (± 0.01)
0.1	1.80×10^{49}	−3.5	−2.7
0.2	1.70×10^{49}	−4.0	−3.0
0.3	3.07×10^{47}	−6.8	−4.2

where A is pre-factor and T_0 is the characteristic temperature coefficient, which takes the form:

$$T_0 = \frac{16\alpha^3}{k\,N(E_{FM})} \tag{6.18}$$

Here, α^{-1} is the localization length, and $N(E_{FM})$ is the density of states at the Fermi level. Low-temperature DC conductivity data with respect to $T^{-0.25}$ are presented in Fig. 6.2d. The experimental data in Fig. 6.2d are fitted to Eq. (6.17). Here, α^{-1} is assumed to be 10 Å [25], which is relevant to the present glassy system with some localized states [25]. The slopes have been computed from the linear best-fit data as shown in Fig. 6.2d. The value of $N(E_{FM})$ has been estimated from Eq. (6.18), which is presented in Table 6.1. The values of $N(E_{FM})$ are found to decrease with composition (x), which show similar nature of DC conductivity. The above-mentioned outcomes convey the facts that decrement in V_2O_5 content in the compositions may be the most important factor for the conduction process in low temperature.

6.5 Conclusion

New Li_2O-doped glassy ceramics have been prepared using melt-quenching route, and their electrical DC conductivity has been studied in wide temperature regime. It is anticipated from the nature of composition that Li^+ conduction mostly contributes to electrical conductivity at high temperature, and Mott's variable-range hopping (VRH) model has been utilized to analyze low-temperature DC conductivity data due to polaron hopping.

References

1. R. Zallen, *Physics of Amorphous Solids* (Wiley, New York, 1983)
2. S. Torquato, Nature **405**, 521 (2000)
3. M.D. Ingram, Phys. Chem. Glasses **28**, 215 (1987)
4. C.A. Angell, Chem. Rev. **90**, 523 (1990)

5. H. Kahnt, J. Non-Cryst. Solids **203**, 225 (1996)
6. K. Otto, Phys. Chem. Glasses **7**, 29 (1966)
7. A. Schtchukarev, W.R. Muller, Z. Phys. Chem. Abst. A **150**, 489 (1930)
8. S.W. Martin, C.A. Angell, J. Non-Cryst. Solids **83**, 185 (1986)
9. W.O. Anderson, D. Stuart, J. Amer. Soc. **37**, 573 (1954)
10. D. McElfresh, D. Howitt, J. Am. Ceram. Soc. **69**, 237 (1986)
11. S. Elliott, J. Non-Cryst. Solids **160**, 29 (1993)
12. C.D. Ravaine, J.L. Souquet, Phys. Chem. Glasses **18**, 27 (1977)
13. J.O. Isard, M. Jagla, K.K. Mallick, J. Physique **43**, 387 (1982)
14. X.Z.R. Yuan, C. Song, H. Wang, J. Jhang, *Electrochmical Impedance Spectroscopy in PEM Fuel Cell: Fundamental and Applications*, XII, 420 p (2010). ISBN: 978-1-84882-845-2
15. A. Acharya, K. Bhattacharya, C.K. Ghosh, S. Bhattacharya, Mater. Lett. **265**, 127438 (2020)
16. G. López-Calzada, Ma.E. Zayas, M. Ceron-Rivera, J. Percino, V.M. Chapela, O. Zelaya-Ángel, S. Jiménez-Sandoval, J. Carmona-Rodriguez, O. Portillo-Moreno, R. Lozada-Morales, J. Non-Cryst. Solids **356**, 374 (2010)
17. A. Ghosh, S. Bhattacharya, D.P. Bhattacharya, A. Ghosh, J. Phys.: Condens. Matter **19**, 106222 (2007)
18. A. Ghosh, S. Bhattacharya, A. Ghosh, J. Phys.: Condens. Matter **21**, 145802 (2009)
19. A. Ghosh, S. Bhattacharya, A. Ghosh, J. Appl. Phys. **101**, 083511 (2007)
20. R.A. Montani, M.A. Frechero, Solid State Ion. **177**, 2911 (2006)
21. T. Holstein, Ann. Phys. **8**, 325 (1959)
22. I.G. Austin, N.F. Mott, Adv. Phys. **18**, 41 (1969)
23. N.F. Mott, Charge transport in non-crystalline semiconductors. Philos. Mag. **19**, 835 (1969)
24. E.A. Davis, N.F. Mott, Phil. Mag. **22**, 903 (1970)
25. M.H.A. Mhareb, S. Hashim, S.K. Ghoshal, Y.S.M. Alajerami, M.A. Saleh, M.M.A. Maqableh, N. Tamchek, Optik **126**, 3638 (2015)

Chapter 7
Frequency-Dependent AC Conductivity of Some Glassy Systems

Sanjib Bhattacharya

Abstract Electrical conductivity of new Li_2O-doped glassy ceramics in wide frequency and temperature regime has been described not only for their applicability in various fields like lithium ion conductors but also for academic interest. Here, this chapter presents "Jonscher's power law model and Almond-West formalism" to interpret mixed conduction process in the present system. It also points that the ratio of power law pre-factor to the exponent ($-\log_{10}$ A/S) indicates temperature independency and strongly composition dependency of present conductors. Based on the transport properties, these glassy ceramics can be treated as suitable candidates for lithium ion battery application with lower lithium content.

Keywords Glass–ceramics · Electrical properties · Li^+ ion migration · Hopping frequency · Frequency exponent

7.1 Introduction

Lithium is considered to be one of the most promising components of rechargeable batteries, which may be applicable to electric vehicles, smart phones and mobile computers [1, 2]. Zheng et al. [3] have pointed out some safety issues, which may arise due to mixing of highly flammable electrolytes. For this reason, traditional electrolytes have been substituted by inorganic solid electrolytes with enough thermal stability, energy density and electrochemical stability [4, 5]. Influence of silver ion concentration on dielectric properties of Li_2O-doped glassy system [6] has been studied extensively, which reveals remarkable contribution of space charge polarization. Attempts have been made to enhance the electrochemical performance of lithium-rich oxide layer material with Mg and La co-doping [7]. Highly resistive lithium-depleted layer [8] has been found in the lithium conducting composites due to the very low mobile ion concentration.

S. Bhattacharya (✉)
UGC-HRDC (Physics), University of North Bengal, Darjeeling, West Bengal 734013, India
e-mail: ddirhrdc@nbu.ac.in; sanjib_ssp@yahoo.co.in

S. Bhattacharya and K. Bhattacharya (eds.), *Lithium Ion Glassy Electrolytes*,
https://doi.org/10.1007/978-981-19-3269-4_7

To shed some light on conductivity spectra and electrical relaxation mechanism in lithium ion conductor, Jonscher's power law model [9, 10] can be employed. In this law, total conductivity of lithium ion conductor can be described as:

$$\sigma(\omega) = \sigma_0 A \omega^s \tag{7.1}$$

where A is the pre-factor, S is the frequency exponent and σ_o is the low frequency or DC conductivity. The conductivity spectra of them can be analysed at various temperatures using Almond-West formalism (power law model) [10],

$$\sigma(\omega) = \sigma_{dc}\left[1 + (\omega/\omega_H)^n\right] \tag{7.2}$$

which is the combination of the DC conductivity (σ_{dc}), hopping frequency (ω_H) and a fractional power law exponent (n). This model can be used to get much information of conducting system [5], which makes it useful tool to interpret electrical conductivity data. A solid-state battery can be constructed from solid-state electrolyte with higher electrical energy density and zero leakage current [1].

New materials hold the key to fundamental advances in energy conversion and storage, both of which are vital in order to meet the challenge of global warming and the finite nature of fossil fuels [2]. Lithium batteries are the systems of choice, offering high energy density, flexible, lightweight design and longer lifespan than comparable battery technologies [3].

Electrical conduction of Li_2O-doped glassy ceramics is expected to closely related to their structure. In the present work, electrical conduction behaviour of Li_2O-doped new glassy ceramics is discussed to explore the scope of new features of lithium ion conductor. Present work is expected to be focused area for research community not only for technological applications but also for academic interest.

7.2 Experimental

Glassy ceramics, xLi_2O–$(1-x)$ $(0.8V_2O_5$–$0.2ZnO)$ with $x = 0.1$, 0.2 and 0.3, have been developed by solid-state reaction. The precursor powders Li_2O, V_2O_5 and ZnO have been thoroughly mixed in proper stoichiometry of the composition. The mixtures are then heated in an alumina crucible in an electric furnace in the temperature range 600°–700 °C for 30 min. Next, the mixtures are allowed to pass through the process of slow cooling during 17 h. The final product is gently crushed to get fine powder. Using a pelletizer at a pressure of 90 kg/cm^2, small pellets (diameter ~20 mm and thickness ~6 mm) of them have been formed. To perform electrical measurement, conducting silver paste has been painted on both sides of the pellets, which are acting as electrodes. The capacitance (C), conductance (G) and dielectric loss tangent (tan δ) of all as-prepared samples have been measured using programmable automatic high-precision LCR metre (HIOKI, model no. 3532-50) at various temperatures in the frequency range 42 Hz–5 MHz. The microstructure of the as-prepared samples is explored by transmission electron microscopic (TEM, JEOL model: JEM-2100 HR) studies. The melting point of alumina is 2072 °C. In the temperature range 600°–700 °C, alumina crucible should not be expected to cause alumina infiltration into the final glass ceramic product. Electron diffraction spectra of the final glass ceramic product also do not exhibit any signature of aluminium.

7.3 Results and Discussion

7.3.1 *Microstructure*

The XRD patterns of the present glassy system are presented in Fig. 7.1a. It is noteworthy from Fig. 7.1a that the XRD patterns of the glassy samples for $x = 0.1$ and 0.2 do not show any sharp peaks; rather they exhibit small peaks, which may explore the nature of poly-crystallinity a little bit. But prominent crystallinity [11] over the amorphous glassy matrices is remarkably observed in the form of sharp peaks in the XRD pattern in Fig. 7.1a for $x = 0.3$. Diffraction peaks in XRD represent the corresponding interplanar spacing (nanostructural periodicity), and sharp peaks represent better crystallinity over the amorphous glass matrices. In this regard, it may be concluded that the as-prepared sample with $x = 0.3$ should represent higher degree of crystallinity and the others, with $x = 0.1$ and 0.2 should exhibit less crystallinity. Formation of nanophases of lithium zinc vanadate ($LiZnVO_4$) with rhombohedral structure has been confirmed from different peaks of Fig. 7.1a and ICDD file no. 38-1332. All the peak positions are indexed except $2\theta = 31°$ and $64°$, because of unavailability of such datasheet. The crystallite size has been estimated from the full width at half maxima (FWHM) of the single diffraction peak in Fig. 7.1a using the Scherer relation [12], $d_c = \frac{0.89\lambda}{\beta \cos\theta}$ where d_c is the crystallite size, λ is the wavelength of X-ray (1.54 Å) radiation, β is the FWHM and θ is the Bragg's diffraction angle. Average crystallite sizes of above-mentioned nanocrystallites with composition (x) have been presented in Fig. 7.1b. It is revealed from Fig. 7.1b that crystallite size increases with composition. This result requires structural modifications of the present system. To validate the results obtained from XRD data, transmission electron micrograph (TEM) of glassy sample, $x = 0.2$ is illustrated in Fig. 7.1c. This micrograph clearly shows the distribution of nanocrystallites of different sizes dispersed in the glassy matrix, and average sizes of them have been estimated. It is also observed that sizes of grains/nanocrystallites dispersed in as-prepared glassy matrices are similar to those obtained from XRD data. The selected area electron diffraction (SAED) pattern for $x = 0.2$ is also displayed Fig. 7.1d. The SAED pattern yields various diffused rings, which are the signatures of amorphous nature [13] of the samples. Superimposition of some shining spots over diffused rings [13] is also observed in Fig. 7.1d, which indicates the presence of certain crystalline plane surfaces of $LiZnVO_4$ nanocrystallites.

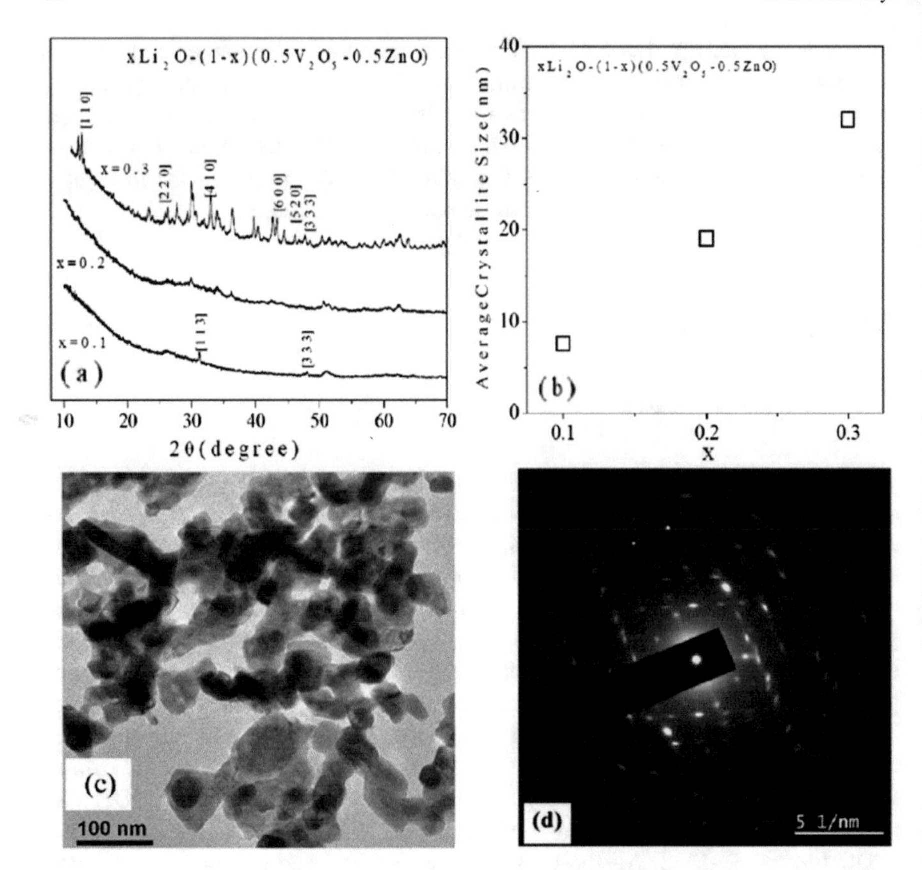

Fig. 7.1 **a** XRD spectra of as-prepared samples; **b** average crystallite sizes with compositions; **c** TEM image for $x = 0.2$ and **d** SAED for $x = 0.2$. Republished with permission from Acharya et al. [14]

7.3.2 Power Law Model and Almond-West Model

Jonscher's power law model [9, 10] and Almond-West formalism (power law model) [10] have already introduced in the introduction section by Eqs. (7.1) and (7.2), respectively. To shed more light on the nature of conductivity spectra as shown in Fig. 7.2a for proper understanding of conduction mechanism due to electronic (polaron) as well as ionic contribution of the lithium-containing glass ceramics, Emin model [15] can be considered, which indicates that there are two sources of the frequency dependence for polaron hopping conductivities. Firstly, the jump rate for a polaron hop increases with applied frequency (except at exceptionally high temperatures) as per Eq. (94) and curve b of Fig. 7 in Ref. [15]. Secondly, the conductivity of carriers confined within spatial regions by disorder rises as the applied frequency is increased [15]. The second effect [15] generates the very low-temperature (<10 K) AC conductivity of charge carriers that hop between especially close pairs of impurity

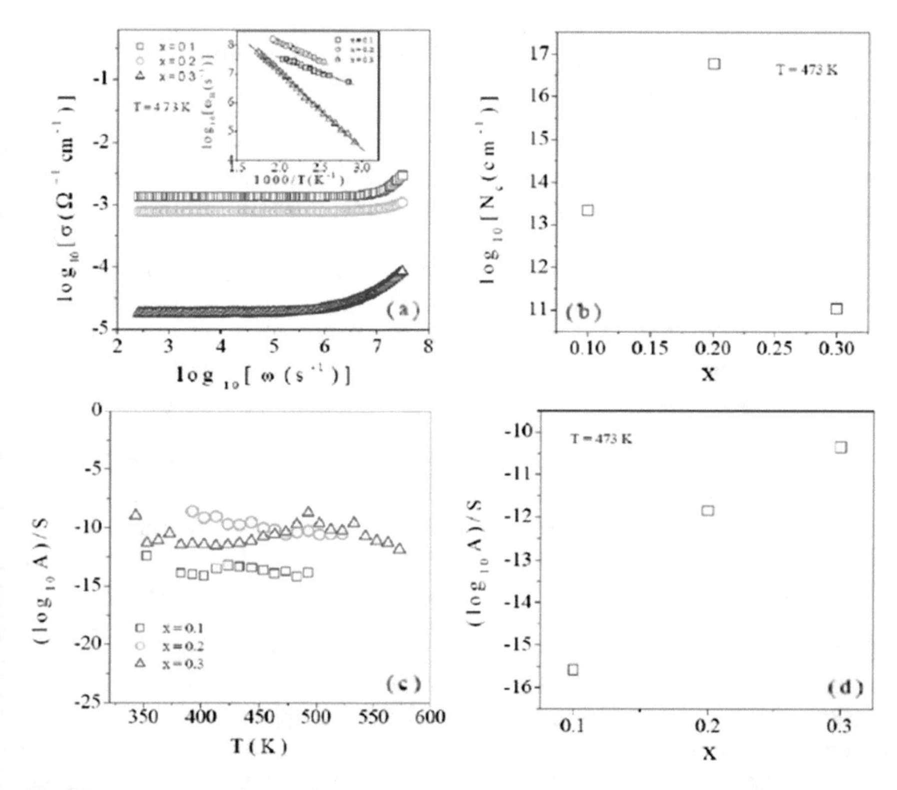

Fig. 7.2 **a** Conductivity spectra for all samples at a fixed temperature; temperature dependency of hopping frequency is shown in the inset; **b** Mobile charge carrier concentration at a fixed temperature; **c** ($\log_{10}$ A)/S with temperature and **d** ($\log_{10}$ A)/S with compositions. Republished with permission from Acharya et al. [14]

states of very lightly doped compensated covalent semiconductors [16]. However, the major issue is to explain of such an effect, which should dominate the high-temperature hopping of high densities of polarons in the present glassy system with $x = 0.1$ and 0.2. To shed some light on this issue, microstructural investigation of the as-prepared samples has been performed. XRD analysis (Fig. 7.1a and b) reveals that various number of small irregular grains of $LiZnVO_4$ of sizes 8–20 nm (less crystallinity) are distributed within the glassy matrix for $x = 0.1$ and 0.2. The smaller grains of $LiZnVO_4$ tend to form agglomerates [17] within the glassy matrix, which should be treated as close pairs of impurity states, which is favourable for both the electron (polaron) transportation and Li^+ penetration.

But lightweight electron (polaron) transportation is expected to dominate over Li^+ penetration for $x = 0.1$ and 0.2. The above-mentioned "second effect" can be realised here at high temperature due to formation of such agglomerated defect states. In addition to this, polaron conduction may be trapped in the small grains in the high-frequency and high-temperature regions. As a consequence, frequency-dependent portion of the conductivity above the frequency-independent portion of the conductivity decreases with increasing Li_2O content up to $x = 0.2$. Recent work [17] reveals that $LiZnVO_4$ nanostructures have been composed of thin nanobelts in a restricted annealing route. $LiZnVO_4$ nanostructures can be used as anode materials for Li^+ intercalation, which exhibited excellent cyclic stability and high rate performance [17]. Figure 7.1a and c exhibits prominent $LiZnVO_4$ nanostructures with irregular shapes for $x = 0.3$. These results indicate that as-prepared sample with $x = 0.3$ does favour Li^+ penetration mostly as diffusion due to short pathways as $LiZnVO_4$ nanostructures form a distributed and correlated structures as confirmed from TEM micrograph of Fig. 7.1c. Owing to the above-mentioned effect, AC conductivity is found to increase over frequency-dependent portion of the conductivity for $x = 0.3$.

The conductivity spectra at 473 K of as-prepared samples are depicted in Fig. 7.2a. It is noted in Fig. 7.2a that lower frequency corresponds to plateau-shaped conductivity (DC conductivity), which is caused by diffusion of mixed charge carriers (Li^+ ions and polaron) [18, 19]. It is also noteworthy from Fig. 7.2a that dispersion starts at higher frequencies, which may be described by a power law [18, 19] as mentioned in Eq. (7.1). Correlated motion [18] of mixed charge carriers may be the possible reason for this dispersion. Ion–polaron interaction may be expected for the outcomes of this correlated motion. The estimated values of power law exponent as mentioned in Eq. (7.1) may reveal the above-mentioned phenomena. Experimental data in Fig. 7.2a has been well-fitted by Eq. (7.2). The values of hopping frequency (ω_H) of the present system have been computed from this fitting. Estimated hopping frequency (ω_H) with reciprocal temperature is presented in the inset of Fig. 7.2a, which also shows thermally activated nature. The values of activation energy for conduction of mixed charge carriers (E_H) corresponding to hopping frequency can be estimated from the linear fit data as shown by solid lines. The estimated values of E_H are also presented in Table 7.1. It is remarkably noted from the inset in Fig. 7.2a that sample with $x = 0.1$ shows intermediate ω_H. Samples with $x = 0.2$ and 0.3 illustrate highest and lowest ω_H, respectively. To explain this anomalous nature of ω_H, concentration of mobile charge carriers (N_c) has been calculated from Nernst–Einstein relation [20]. Composition dependent N_c at a fixed temperature has been presented in Fig. 7.2b, which shows that initially, N_c starts to increase with x and then decreases with x. N_c may be related with conductivity [21] as $\sigma_{dc} = N_c q\mu$, where q is the charge carrier and μ is the mobility of charge carrier. σ_{dc} is found to decrease with Li_2O content in Fig. 7.1c and σ_{dc} becomes highest for $x = 0.1$. But $x = 0.1$ corresponds to lower

Table 7.1 The estimated values of interplanar spacings (*d* values), activation energy corresponding to hopping frequency (EH) and fixed frequency AC conductivity of *x*Li2O–(1−*x*)(0.5V2O5–0.5ZnO) glass–ceramics. Estimated errors of measurements are mentioned

x	d (nm) (± 0.01)	E_H (eV) (± 0.001)	$\log_{10}[\sigma'(\Omega^{-1}\ cm^{-1})]$ at 4.8 MHz (± 0.01)
0.1	2.88	0.30	−2.7
0.2	1.89	0.35	−3.0
0.3	3.50, 2.68, 1.89	0.52	−4.2

value of N_c as well as ω_H. These results directly indicate higher values of mobility of charge carrier (μ) to validate the above-mentioned relation of σ_{dc}. Since polaron is supposed to be lighter than lithium ion, highest σ_{dc} for $x = 0.1$ can be anticipated mostly due to polaron hopping and partly due to conduction of Li^+. In this context, it may be concluded that sample with $x = 0.2$ exhibits lower mobility of charge carrier (μ), because of mostly due to conduction of Li^+ and partly due to polaron hopping. Conduction of Li^+ dominates for $x = 0.3$ with lower μ.

Power law pre-factor (A) can be employed to explain temperature and composition dependency of lithium ion conductor [22, 23]. Pre-factor (A) can be estimated from Eq. (7.1) as: $A = \sigma_{dc} \times \omega_H{}^{-S}$, which is already computed by another groups [22, 23] to establish composition dependency of $-\log_{10}$ A/S [9]. Variation of $-\log_{10}$ A/S of the present system has been presented in Fig. 7.2c with temperature, which exhibits a constant value for the entire temperature window. This result directly suggests that the temperature advancement of $\log_{10}A$ is comparative to the temperature advancement of S and the conduction process may be anticipated as conduction of mixed charge carriers [18, 19]. But the present system is strongly composition dependent, which is observed from Fig. 7.2d. It is also noted from Table 7.1 that the values of n, obtained from Eq. (7.2) and the values of S, obtained from Eq. (7.1) are not same due to the presence of mixed charge carriers. Estimated values of S and n may be related to the three-dimensional [18, 19] motions of charge carriers.

7.3.3 Others

Our extensive study [24] on three glassy ceramics, $0.1Li_2O$–$0.9(0.5\ SeO_2$–$0.5\ P_2O_5)$ (base), $0.1Li_2O$–$0.9\ (0.4\ SeO_2$–$0.1Nd_2O_3$–$0.5\ P_2O_5)$ (base doped with Nd_2O_3) and $0.1Li_2O$–$0.9\ (0.4\ SeO_2$–$0.1MoO_3$–$0.5\ P_2O_5)$ (base doped with MoO_3) revels that base sample exhibits highest AC conductivity and MoO_3-doped as-prepared sample exhibits lowest AC conductivity, which is similar to the nature of variation of DC conductivity. As SeO_2 in the composition actively participates in the bonding as well as in the formation of larger cluster nanoassembly [25], Li^+ ions can easily migrate via cluster hopping. Doping of Nd_2O_3 in the base may partially transform SeO_2 cluster nanoassembly into SeO_2 chain structure [26]. Apart from that, Nd_2O_3 may take part in the partial covalent bonding [26] of the resultant glassy ceramics. This may lead to decrease of conductivity level as the network of glass structure offers more resistive pathways for migration of Li^+ ions. Nature of MoO_3-doped glassy sample [27] exhibits thermodynamically stable MoO_3 of orthorhombic structure with corner-shared MoO_6 octahedra. Because of small ionic radius (1.52 Å), Li^+ ions must be surrounded by the strong electronegative $MoO_4{}^{2-}$ entities [27], which makes Li^+ ions less mobile. Variation of conductivity may be explained from structural behaviour of as-prepared samples. For base sample (higher SeO_2 content), the isolated $Se = O$ bonds should form chain [28], which may oppose migration of Li^{+2} ions. Likewise, Nd_2O_3-doped and MoO_3-doped samples must exhibit higher conductivities, because they contain less amount of SeO_2. But experimental evidence reveals different scenario. Base sample exhibits highest conductivity. This result directly indicates that P_2O_5 plays important role in this issue. Phosphorous is very reactive, which may form a variety of P–Se heterocycles [29] in the base sample. This process may finally release some complexes and polynuclear clusters of SeO_3 and SeO_4, which is favourable for conduction of Li^{+2} ions. As a consequent, conductivity of base sample is high. Phosphorous can adsorb neodymium [30] in the composition, which may resist more cluster formation of SeO_3 and SeO_4. So the formation of chains of SeO_3 and SeO_4 in the Nd_2O_3-doped sample may be the possible reason for less electrical conductivity. In MoO_3-doped sample, multiple bonding between transition metal (Mo) and phosphorus may form a distributed complexes [14], which is responsible for restricted ion motion in the composition. This may directly indicate the reason for very less electrical conductivity for MoO_3-doped sample.

7.4 Conclusion

AC conductivity data of present glassy system containing Li_2O reveals that Li^+ conduction mostly contributes to electrical conductivity at high temperature. Power law pre-factor (A) has been employed to explain temperature and composition dependency of lithium ion conductor. The power law exponent (S) is not observed to be same due to the presence of mixed charge carriers in the present system. It is remarkably noted that sample with $x = 0.1$ shows intermediate hopping frequency (ω_H). Samples with $x = 0.2$ and 0.3 illustrate highest and lowest ω_H, respectively. This anomalous nature of ω_H has been described with the help of mobile charge carrier's concentration.

Acknowledgements The financial assistance for the work by the Council of Scientific and Industrial Research (CSIR), India, via sanction No. 03(1411)/17/EMR-II is thankfully acknowledged. A. Acharya, C. K. Ghosh and K. Bhattacharya are also thankfully acknowledged for their research support.

References

1. K. Takada, J. Power Sources **394**, 74 (2018)
2. H. Pan, S. Zhang, J. Chen, M. Gao, Y. Liu, T. Zhu, Y. Jiang, Mol. Syst. Des. Eng. **3**, 748 (2018)
3. F. Zheng, M. Kotobuki, S. Song, M.O. Lai, L. Lu, J. Power Sources **389**, 198 (2018)
4. X. Yao, B. Huang, J. Yin, G. Peng, Z. Huang, C. Gao, D. Liu, X. Xu, Chin. Phys. B **25**, 018802 (2016)
5. E. Zanotto, Am. Ceram. Soc. Bull. **89**, 19 (2010)
6. V. Prasad, L. Pavić, A. Moguš-Milanković, A. Siva Sesha Reddy, Y. Gandhi, V. Ravi Kumar, G. Naga Raju, N. Veeraiah, J. Alloys Compd. **773**, 654 (2019)
7. H. Seo, H. Kim, K. Kim, H. Choi, J. Kim, J. Alloys Compd. **782**, 525 (2019)
8. K. Takada, T. Ohno, N. Ohta, T. Ohnishi, Y. Tanaka, ACS Energy Lett. **3**, 98 (2018)

9. A.K. Jonscher, *Dielectric Relaxation in Solids* (Chelsea Dielectrics) (London, 1983); A.K. Jonscher, Nature **267**, 673 (1977)
10. D.P. Almond, A.R. West, Nature **306**, 456 (1983); D.P. Almond, G.K. Duncan, A.R. West, Solid State Ion. **8**, 159 (1983)
11. P.W. Jaschin, K.B.R. Varma, J. Non-Cryst. Solids **434**, 41 (2016)
12. S. Singhal, J. Kaur, T. Namgyal, R. Sharma, Phys B **407**, 1223 (2012)
13. S. Bhattacharya, A. Ghosh, J. Am. Ceram. Soc. **91**, 753 (2008)
14. A. Acharya, K. Bhattacharya, C.K. Ghosh, A.N. Biswas, S. Bhattacharya, Mater. Sci. Eng. B **260**, 114612 (2020)
15. D. Emin, Adv. Phys. **24**, 305 (1975)
16. M. Pollak, T.H. Geballe, Phys. Rev. **122**, 1742 (1961)
17. L. Zeng, X. Huang, C. Zheng, Q. Qian, Q. Chen, M. Wei, Dalton Trans. **44**, 7967 (2015)
18. S.R. Elliott, *Physics of Amorphous Materials* (John Wiley and Sons, Incorporated, 1986). ISBN 0470204729, 9780470204726
19. N.F. Mott, E.A. Davis, *Electronic Processes in Non-Crystalline Materials* (Clarendon Press, Oxford, 1979)
20. S. Bhattacharya, A.K. Bar, D. Roy, M.P.F. Graca, M.A. Valente, Adv. Sci. Lett. **16**, 399 (2012)
21. M.M. Ahmad, Phys. Rev. B **72**, 174303 (2005)
22. A.N. Papathanassiou, Mater. Let. **59**, 1634 (2005)
23. J.L. Ndeugueu, M. Aniya, J. Phys. Soc. Jpn. **79**, 72 (2010)
24. S. Bhattacharya et al., Phys. B **546**, 10 (2018)
25. A.K. Bar, K. Bhattacharya, R. Kundu, D. Roy, S. Bhattacharya, Mater. Chem. Phys. **199**, 322 (2017)
26. H.M. Xiong, X. Zhao, J.S. Chen, J. Phys. Chem. B **105**, 10169 (2001)
27. S. Bhattacharya, A. Ghosh, Appl. Phys. Lett. **88**, 133122 (2006)
28. N. Gupta, A. Dalvi, Indian J. Pure Appl. Phys. **51**, 328 (2013)
29. G. Hua, J. Derek Woollins, Angew. Chem. Int. Ed. **48**, 1368 (2009)
30. H.J. Park, L.L. Tavlarides, Ind. Eng. Chem. Res. **49**, 12567 (2010)
31. M. Takashima, S. Yonezawa, Y. Ukuma, J. Fluorine Chem. **87**, 229 (1998)
32. A.H. Cowley, A.R. Barron, ACC Chem. Res. **21**, 81 (1988)

Chapter 8
Dielectric Properties and Analysis of Some Li-Doped Glassy Systems

Amartya Acharya, Koyel Bhattacharya, Chandan Kr Ghosh, and Sanjib Bhattacharya

Abstract A series of new glass–ceramic samples containing Li_2O has been prepared to explore their dielectric properties in the frequency range 42 Hz–5 MHz and at several temperatures. XRD patterns and TEM micrographs of them reveal the formation of different types of nanocrystallites, dispersed in amorphous glassy matrices. The imaginary component of the electric modulus has been studied to reveal the temperature dependency of stretched coefficient (β). The present study also exhibits non-Debye type conductivity relaxation process. The estimated relaxation time (τ_R) shows thermally activated nature. The values of relaxation activation energy (E_R) of as-prepared samples indicate that the charge carriers must overcome the energy barrier during the relaxation process. Electric modulus scaling spectra indicate that the dynamical relaxation process in the present glassy system is independent of temperature, but depends on composition.

Keywords Li^+ ion-conducting glassy nanocomposites · Relaxation time · X-ray diffraction and TEM · Electric modulus · Li^+ ion relaxation

8.1 Introduction

Glass nanocomposites containing lithium ions are of great interest due to their applications in various wings such as rechargeable batteries, sensors and electrolytes. [1, 2]. To exploit their applications in such fields, some safety issues [3] should be taken into account because of mixing of highly flammable materials to reach the goal. This

A. Acharya · S. Bhattacharya (✉)
UGC-HRDC (Physics), University of North Bengal, Darjeeling, West Bengal 734013, India
e-mail: ddirhrdc@nbu.ac.in; sanjib_ssp@yahoo.co.in

K. Bhattacharya
Department of Physics, Kalipada Ghosh Tarai Mahavidyalaya, Bagdogra, Darjeeling, West Bengal 734014, India

A. Acharya · C. K. Ghosh
Department of Electronics and Communication Engineering, Dr. B. C. Roy Engineering College, Durgapur, West Bengal 713026, India

S. Bhattacharya and K. Bhattacharya (eds.), *Lithium Ion Glassy Electrolytes*,
https://doi.org/10.1007/978-981-19-3269-4_8

idea should compel us to replace traditional electrolytes by inorganic solid electrolytes with enough thermal stability, energy density and electrochemical stability [4, 5]. Vanadium pentoxide (V_2O_5) has different valence states like V^{3+}, V^{4+}, V^{5+} and V_2O_5 has well-known structure consist of VO_5 pyramids and VO_4 tetrahedral as the main coordination for V-atoms [6], while zinc oxide (ZnO) has structure composed of ZnO_4 tetrahedra and ZnO_6 octahedra as the main coordination for Zn-atom [7]. The dielectric constant and dielectric loss are the most critical parameters to design microelectronic equipment [8]. Additionally, researchers are concerned about the microscopic mechanisms responsible for dielectric relaxation, as well as the study of dielectric loss factor as a function of temperature and frequency, which is identified as one of the most suitable and delicate methods of reviewing such glassy structure [9, 10].

To study dielectric properties of composite materials real ($\varepsilon^/$) and imaginary ($\varepsilon^{//}$) parts of complex dielectric permittivity (ε) are supposed to be essential. They may be represented as:

$$\varepsilon^/ = \frac{Ct}{\varepsilon_0 A} \text{ and } \varepsilon^{//} = \varepsilon^/ \tan\delta \tag{8.1}$$

where ε_0 is permittivity of free space, t and A are thickness and area of the glassy samples, respectively.

Again imaginary part of complex permittivity ($\varepsilon^{//}$) has been investigated as per Guintini's model on charge carriers hopping over a potential barrier in the charged defect sites [10, 11]. The outcomes of this model [12] can be summarized as:

$$\varepsilon^{//} = A\omega^m \tag{8.2}$$

and,

$$m = -\frac{4 * K_B T}{W_M} \tag{8.3}$$

where A is a constant, m is the temperature dependent frequency power parameter, and W_M is the maximum barrier height that is the required energy to transport the electron from one site to the next.

To explore dielectric relaxation behaviour of the present glassy system, electrical modulus formalism [13, 14] has been employed. The reciprocal of complex permittivity has been designated as complex electric modulus [13, 14]:

$$M^* = \frac{1}{\varepsilon^*} = M^/ + jM^{//} = \frac{\varepsilon^/}{\left(\varepsilon^/\right)^2 + \left(\varepsilon^{//}\right)^2} + j\frac{\varepsilon^{//}}{\left(\varepsilon^/\right)^2 + \left(\varepsilon^{//}\right)^2} \tag{8.4}$$

where $M^{/}$, $M^{//}$ and $\varepsilon^{/}$, $\varepsilon^{//}$ are real and imaginary parts of complex electric modulus and dielectric permittivity, respectively. The advantage of considering it is to explore the microscopic property of relaxation phenomena [13, 14].

In this chapter, the dielectric properties have been addressed in terms of electrical modulus spectra of Li_2O containing glassy nanocomposites. The results must open the way to the development, usage and study of significant dielectric properties of new type of glass nanocomposites not only for their unusual electrical properties for the purpose of various device applications, but also for academic interest.

8.2 Experimental

Glass nanocomposites, xLi$_2$O–(1−x) (0.8V$_2$O$_5$–0.1ZnO) with $x = 0.1$, 0.2 and 0.3, have been developed from the 99% pure precursors; lithium oxide (Li_2O), vanadium pentoxide (V_2O_5) and zinc oxide (ZnO) via melt quenching process [15]. The proper amounts of reagent grade chemicals have been stoichiometrically mixed and then melted in an electric furnace in the temperature window from 600° to 700 °C. For dielectric measurement, conducting silver paste has been painted on both sides of the samples, acting as an electrode. The capacitance (C), conductance (G) and dielectric loss tangent (tan δ) of the as-prepared samples have been measured using HIOKI (model no. 3532-50) made high-precision LCR meter at different temperatures in the frequency range 42 Hz–5 MHz.

8.3 Results and Discussion

8.3.1 Microstructure

The XRD patterns of the present glassy system are presented in Fig. 8.1a. It is noteworthy from Fig. 8.1a that the XRD patterns of the glassy samples for $x = 0.1$ and 0.2 do not show any sharp peaks; rather they exhibit small peaks. This nature of X-ray diffractograms may explore the nature of polycrystallinity. Crystallinity [16] over the amorphous glassy matrices are remarkably observed in Fig. 8.1a for $x = 0.3$. Different peaks in Fig. 8.1a exhibit the formation of lithium zinc vanadate ($LiZnVO_4$) nanophases with rhombohedral structure, which has been confirmed from ICDD file no 38-1332. All the peak positions are properly indexed except $2\theta =$ 31° and 64°, because of unavailability of such datasheet. The average crystallite size has been estimated from the full width at half maxima (FWHM) of the single diffraction peak in Fig. 8.1a using the Scherer relation [17], $d_c = \frac{0.89\lambda}{\beta\cos\theta}$ where d_c is the crystallite size, λ is the wavelength of X-ray (1.54 Å) radiation, β is the FWHM, and θ is the Bragg's diffraction angle. Average crystallite sizes of above-mentioned nanocrystallites with composition (x) have been shown in Fig. 8.1b. Figure 8.1b

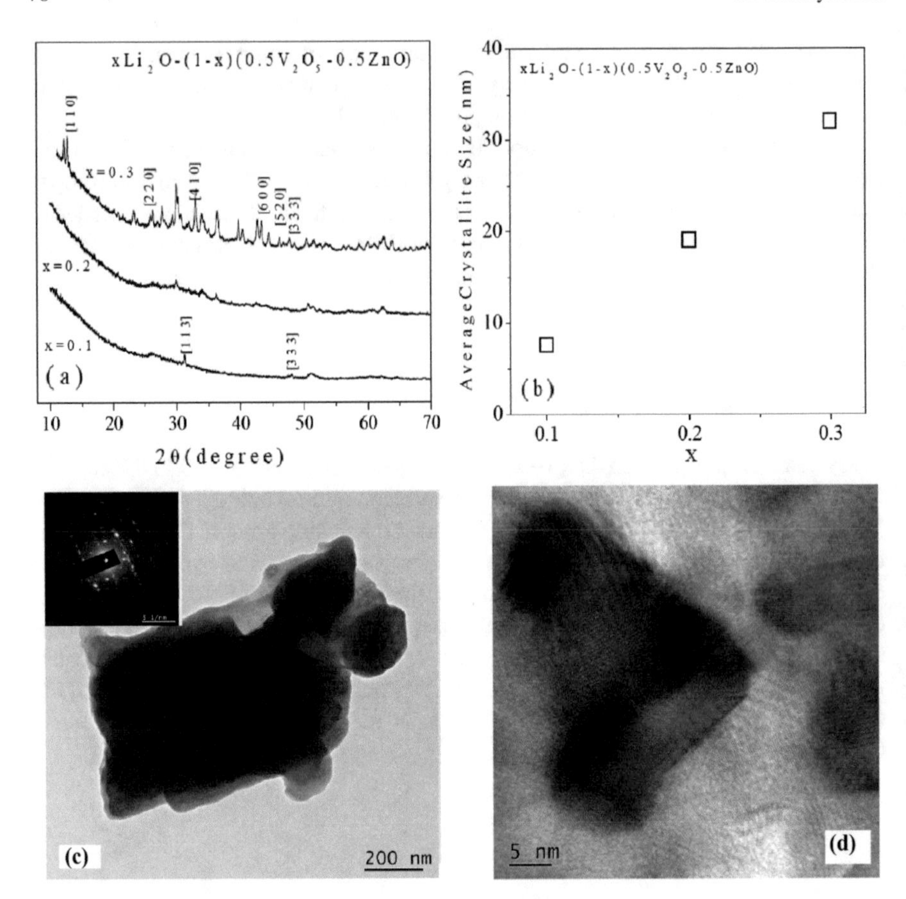

Fig. 8.1 **a** XRD spectra of as-prepared samples; **b** average crystallite sizes with compositions; **c** TEM image for $x = 0.2$ with resolution 200 nm; selected area diffraction (SAED) pattern for $x = 0.2$ in the inset and **d** HRTEM image for $x = 0.2$

clearly reveals that crystallite size increases with composition. This result requires structural alterations of the present system. To establish microstructural informations of XRD data, transmission electron micrographs (TEM) for $x = 0.2$ with different resolutions are presented in Fig. 8.1c and d, respectively. This micrograph clearly provides the information about the distribution of nanocrystallites of different sizes, dispersed in the glassy matrix, which are essentially helpful to estimate average sizes of crystallites. It is also observed that the estimated sizes of grains/nanocrystallites, dispersed in as-prepared glassy matrices are almost similar to those obtained from XRD data. The selected area electron diffraction (SAED) pattern for $x = 0.2$ is also included in the inset of Fig. 8.1c. The SAED pattern gives birth to various diffused rings, which are the signatures of amorphous nature [18] of the as-prepared samples. Some shining spots on the diffused rings [18] are also noted in SAED pattern, which indicate the existence of certain crystalline plane surfaces of $LiZnVO_4$ nanocrystallites.

8.3.2 Study of Dielectric Constant

Experimental data on electrical measurements may impel to shed some light on electrical relaxation process [10, 13, 14]. The real and imaginary parts of complex dielectric permittivity (ε^*) have been computed from Eq. (8.1). Figures 8.2a and b show the frequency dependent dielectric constant, $\varepsilon^/$ and $\varepsilon^{//}$, respectively, for $x = 0.3$ at several temperatures. Figure 8.2a reveals gradual decrease in dielectric constant ($\varepsilon^/$) with frequency, and finally, it would reach to the almost constant limiting value, which can be credited to oscillations of free dipoles in an alternating electric field [10, 13, 14]. Space charge polarization may be responsible for such distribution at lower frequencies for various temperatures [10, 13, 14]. As the frequency rises, the dipoles could not rotate as much in commencing to lag those of the electric field [10, 13, 14]. As the frequency is further increased, the dipoles could not pursue the field completely, and this result may directly indicate to pause orientation polarization. In

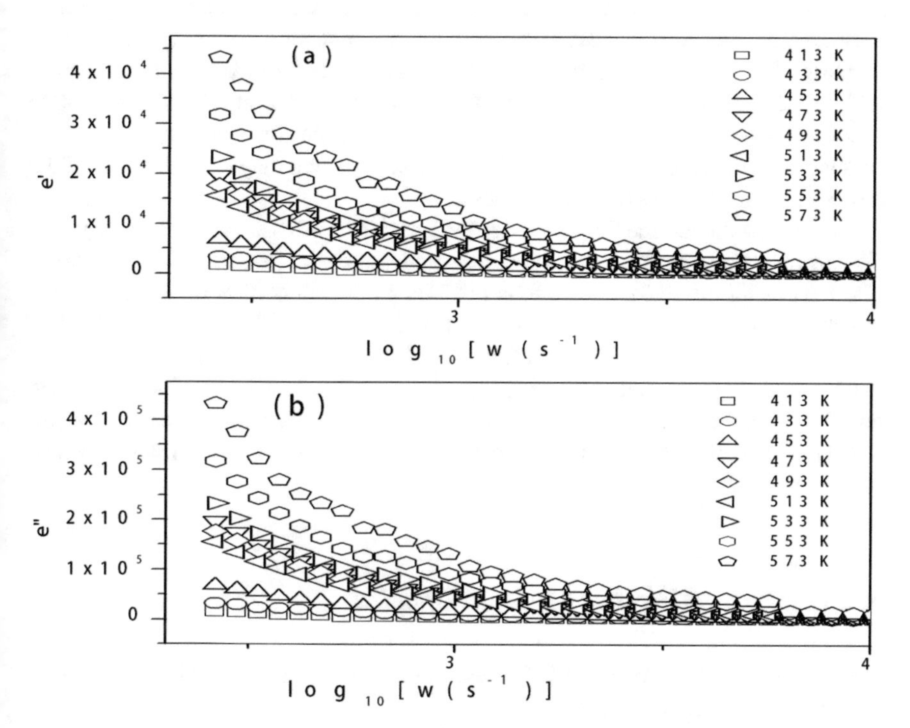

Fig. 8.2 Frequency dependent dielectric constant, **a** $\varepsilon^/$ and **b** $\varepsilon^{//}$, respectively of the glassy composite, $x = 0.3$ at several temperatures

Table 8.1 Maximum barrier height (W_m), dielectric loss (tan δ) at fixed frequency 3 MHz and at temperature 533 K and the activation energy (E_R) associated with the relaxation process. The errors from fitting parameter are also included

x	tan δ (±0.01)	W_m (±0.01) (eV)	E_R (±0.02) (eV)
0.1	7.412	0.051	0.051
0.2	8.211	0.072	0.071
0.3	8.643	0.081	0.080

consequence of the space charge effect or interfacial polarization, the $\varepsilon^{/}(\omega)$ values are expected to drop to a constant value at higher frequencies [10, 13, 14]. The increase of real part of dielectric constant ($\varepsilon^{/}$) with temperature may be related to the reduction in bond energies [19]. The above-mentioned results directly indicate thermally activated dipolar polarization, whose outcomes may be the deteriorations of inter-molecular forces to enhance in orientation vibrations and the formation of thermal agitation to give rise disturbance in orientation vibrations. As a consequence, $\varepsilon^{/}(\omega)$ is found to be almost same. Figure 8.2b depicts that the dielectric constant ($\varepsilon^{//}$) rises with temperature, similar to the phenomena seen in Fig. 8.2a. $\varepsilon^{//}$ is directly proportional to dielectric loss, which may be represented as tan $\delta = \varepsilon^{//}/\varepsilon^{/}$. Estimated values of tan δ (loss tangent) are presented in Table 8.1. Enhancement in temperature may directly lead to the consequent increase of conduction losses, which leads to increase of $\varepsilon^{//}$ with temperature. Figure 8.2b shows that $\varepsilon^{//}$ decreases with frequency. At lower frequencies, $\varepsilon^{//}$ increases owing to dipole polarization [10]. The migration of ions may be predicted as the foremost reason for such changes in $\varepsilon^{//}$ at the lower frequencies [10]. The ion vibrations in the present system may be collapsed at high frequency, which shows constant values of $\varepsilon^{//}$.

Figure 8.3a depicts the frequency dependent $\varepsilon^{//}$ for $x = 0.3$ at various temperatures. The power parameter (m) has been computed from the slopes of straight line fits in Fig. 8.3a using Eq. (8.3). Figure 8.3b exhibits the variation of m with temperature, which shows thermally activated nature. It is also observed from Fig. 8.3b that that m decreases with temperature for all values of x except $x = 0.2$ with an anomalous nature. It needs more study in the near future. From the slope of the plot in Fig. 8.3b and Eq. (8.3), the values of W_M have been estimated and listed in Table 8.1. It is also noted that dielectric constant and dielectric loss of a material at lower frequency tend to increase with temperature.

8.3.3 *Study of Electric Modulus Spectra*

The complex electric modulus (M*), seen in Eq. (8.4), can be used to explore the phenomena of space charge relaxation [10, 13, 14]. The concept of complex electric modulus $M^*(\omega)$ spectra as a function of AC conductivity been successfully applied

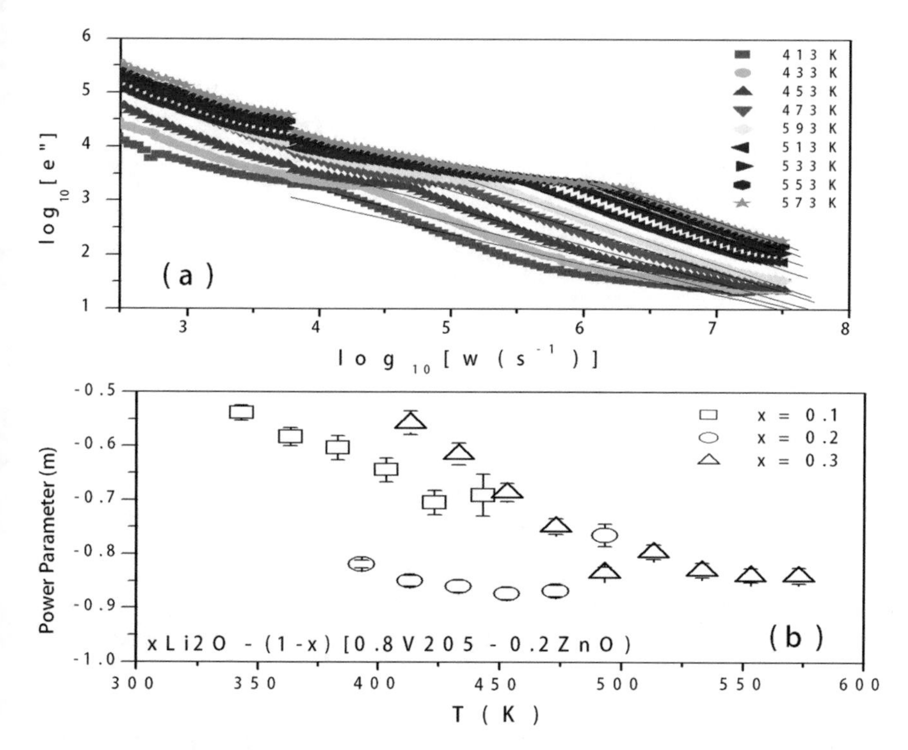

Fig. 8.3 **a** Plot of log $\varepsilon^{//}$ versus log ω of the glassy sample, $x = 0.3$ at various temperatures; **b** variation of power parameter (m) with temperature

here to explain dielectric relaxation of as-prepared samples as it explores the electrical properties of a sample by suppressing the electrode polarization effects [10]. Figures 8.4a and b present the frequency dependent $M^{/}$ and $M^{//}$ values at various temperatures for $x = 0.3$, respectively. Similar nature is noticed for other samples under study. It is observed in Fig. 8.4a that at the lower frequency, $M^{/}$ approaches to zero due to lack of restoring forces of mobile ions [10, 13, 14]. At higher frequencies, $M^{/}$ shows dispersion and finally attains a maximum value due to electrical relaxation of charge carriers, which corresponds to $(M_{\infty}) = (\varepsilon_{\infty})^{-1}$ [10, 13, 14]. This result suggests to the existence of relaxation time that goes along with a loss peak as shown in Fig. 8.4b. Decrement in $M^{/}$ with temperature may be inferred that the conduction mechanism in the present glassy system might be expected to happen due to short-range mobility of charge carriers [10, 13, 14]. Above discussion suggests that the molecular dipoles and the charge carrier orientation may be developed at higher temperatures, leading to the enhancement of charge carrier mobility with temperature.

It is evident from Fig. 8.4b that at lower frequencies $M^{//}$ is low as large value of parallel capacitance has been developed due to contribution of electrode polarization effect [10, 13, 14] via accumulation of huge quantity of charge carriers at the

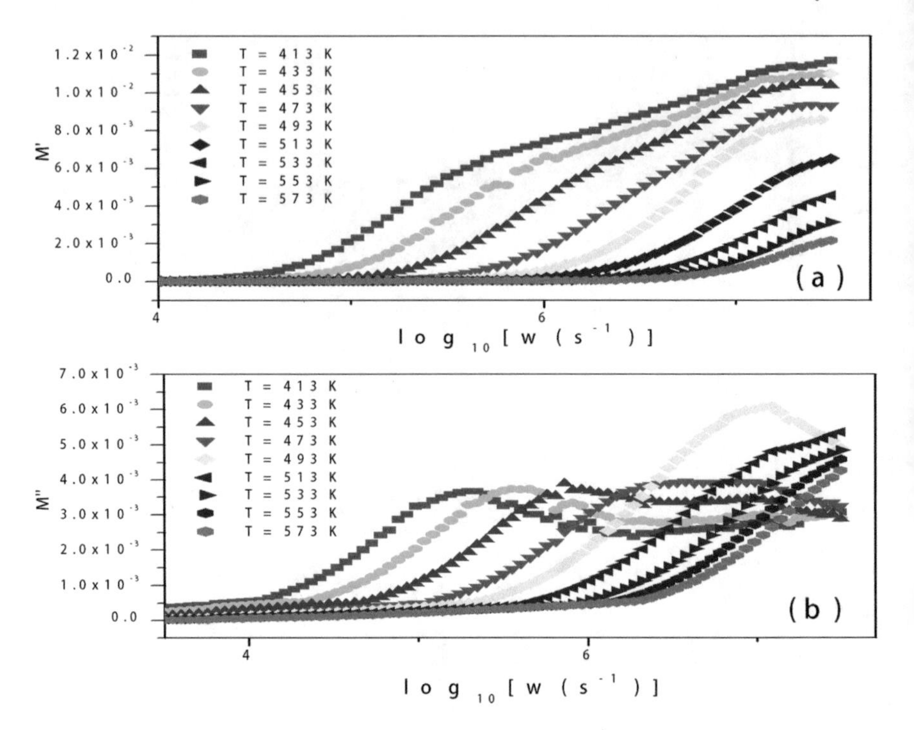

Fig. 8.4 Frequency dependent **a** real part of electric modulus, $M^{/}$ and **b** imaginary part of electric modus, $M^{//}$ at various temperatures for $x = 0.3$

electrode-glass interface [10, 13, 14]. It is clear from Fig. 8.4b that $M^{//}$ exhibits a distinct peak ($\omega_{\max}$), which has been developed due to relaxation dynamics of charge carriers [10, 13, 14]. This relaxation peak is shifted toward higher frequencies with temperature, which shows their thermal activated nature. The charge carriers are expected to move as they are activated thermally, which may reduce relaxation time and therefore, may increase the rate of relaxation frequency. Below the peak relaxation frequency, the conduction mechanism may be caused due to hopping of charge carriers over long distances. Moreover, above the peak relaxation frequency (i.e., above $\omega_{\max}$), the conduction mechanism may be interpreted by the localized motion of charge carriers over short distances. Thus, the relaxation peak signifies the change over from long-range hopping conduction to the short-range localized motion of charge carriers. M^{*} can also be formulated as Fourier transform of relaxation function φ(t):

$$M^{*} = M_{\infty}\left[1 - \int_{0}^{\infty} \exp(-\omega t)\left(\frac{\mathrm{d}\varphi}{\mathrm{d}t}\right)\mathrm{d}t\right] \tag{8.5}$$

Here, $\varphi(t)$ is the time evolution function of the electric field inside the materials [10, 13, 14], usually known as the Kohlrausch–Williams–Watts (KWW) function:

$$\varphi(t) = \exp\left[-\left(\frac{t}{\tau_m}\right)^{\beta}\right] \tag{8.6}$$

where $0 < \beta < 1$, τ_m represents conductivity relaxation time and the exponent β is KWW stretched coefficient [10, 13, 14]. Bergman [20] successfully modified KWW function to introduce new form of $M^{//}$ as per requirement of its nature as:

$$M^{//} = \frac{M^{//}_{\max}}{(1-\beta) + \frac{\beta}{1+\beta}\left[\beta\left(\frac{\omega_{\max}}{\omega}\right) + \left(\frac{\omega}{\omega_{\max}}\right)^{\beta}\right]} \tag{8.7}$$

where $M^{//}{}_{\max}$ is the maximum value of $M^{//}$ and $\omega_{\max}$ represents equivalent maximum angular frequency and β is the KWW stretched coefficient. The values of β have been estimated from fitting of data in Fig. 8.4b using Eq. (8.7). The variation of β with temperature has been presented in Fig. 8.5a, which exhibits decrease in β with temperature. It is observed from Fig. 8.5a that the estimated value of β is less than 1,

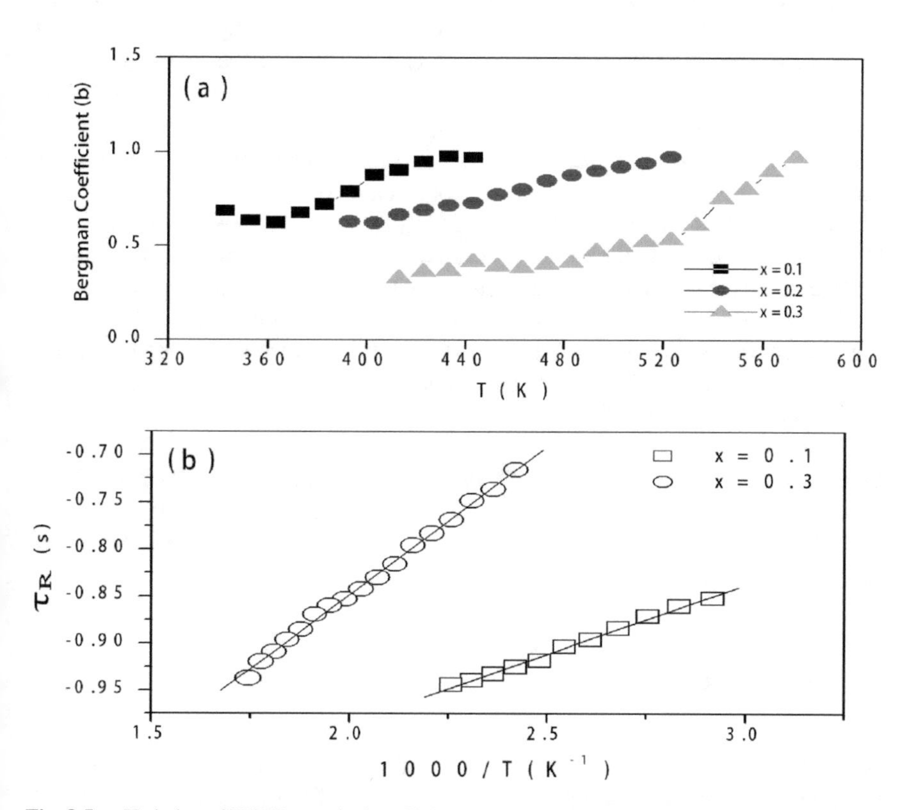

Fig. 8.5 **a** Variation of KWW stretched coefficient (β) with temperature of all as-prepared samples; **b** Rrelaxation time (τ_R) with reciprocal temperature

which confirms that the relaxation process is non-Debye type [10, 13, 14, 21]. The relaxation frequency ($f_{\max}$) corresponding to maximum value of $M^{//}$ in Fig. 8.4b has been obtained from the fitting using Eq. (8.7). In this approach, the relaxation time (τ_R) has been computed from the relation [21]:

$$\tau_R = \frac{1}{2\pi f_{\max}} = \frac{1}{\omega_{\max}} \tag{8.8}$$

The activation energy (E_R) corresponding to τ_R has been estimated using the relation [22]:

$$\tau_R = \tau_0 \exp\left(-\frac{E_R}{K_B T}\right) \tag{8.9}$$

where τ_0 is a pre-exponential factor, K_B is the Boltzmann constant, E_R is the corresponding activation energy and T is the absolute temperature. The temperature dependency of τ_R is depicted in Fig. 8.5b. E_R has been estimated from the best-fitted straight-line plots in Fig. 8.5b. Figure 8.5b also confirms that the relaxation time (τ_R) decreases with temperature. Here, it is also noted from Fig. 8.5a that KWW stretched coefficient (β) [14, 15] increases with temperature, which may directly indicate that more electrical stress has been imposed to the present system. So, the electrical relaxation time increases with composition, which is evident from Fig. 8.5b. It is also observed from Table 8.1 that W_m and E_R are comparable, which indicate that lithium ion in the present system can overcome potential barrier with electrical relaxation process alone. So, slight triggering by external reagent should change electrical relaxation greatly, which indicates strongly composition dependent nature.

To explore the nature of electrical relaxation process of the present glassy system, scaling process of $M^{//}$ has been employed. Temperature scaling of $M^{//}$ [15] is presented in Fig. 8.6a at various temperatures for $x = 0.1$. In this scaling process [15], $M^{//}$ axis is divided by $M^{//}{}_{\max}$, while the frequency axis is divided by the conductivity relaxation frequency ($\omega_{\max}$). A perfect overlap of all $M^{//}$ spectra at various temperatures is observed in Fig. 8.6a, which indicates temperature independent relaxation process. Composition scaling of $M^{//}$ spectra is presented in Fig. 8.6b at 433 K, which shows non-overlapping curves for all compositions. It may be concluded from the scaling phenomena that the electrical relaxation process is independent of temperature, but depends on composition.

8.4 Conclusion

Study of dielectric property of some lithium zinc vanadate glassy nanocomposites using electric modulus formalism in a wide range of temperature and frequency reveals that conductivity relaxation process is non-Debye type and the values of

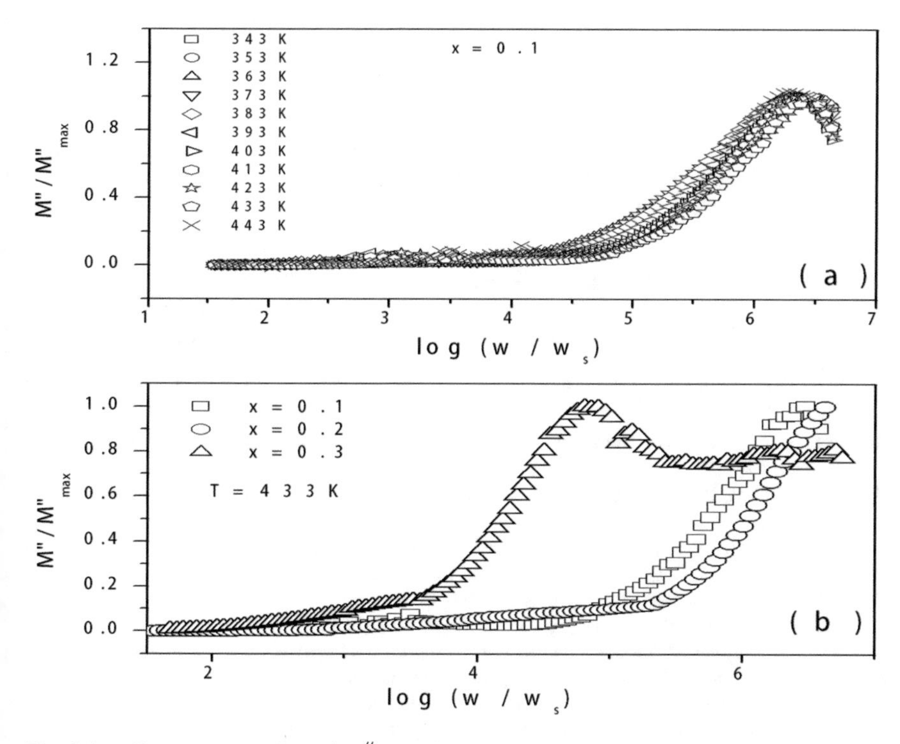

Fig. 8.6 **a** Temperature scaling of $M^{//}$ at various temperatures for $x = 0.1$; **b** the composition scaling spectra of $M^{//}$ at 433 K

the relaxation time (τ_R) decrease with temperature. The dielectric constant ($\varepsilon^{/}$) and dielectric loss ($\varepsilon^{//}$) values are found to decrease with frequency and finally attain a nearly constant value. $\varepsilon^{/}$ and $\varepsilon^{//}$ are also show thermally activated nature. The maximum barrier height (W_M) has been estimated at various temperatures. $M^{//}$ spectra have been fitted well using Bergman's function, and stretched coefficient (β) values have been found to decrease with temperature. The values of relaxation activation energy (E_R) suggest that the charge carriers must overcome the energy barrier during the relaxation process. Perfect overlap of temperature scaling spectra of $M^{//}$ of all the glass nanocomposites at various temperatures indicate temperature independent relaxation process. However, imperfect overlap of composition scaling of $M^{//}$ spectra at a fixed temperature reveal composition dependence of relaxation process of charge carriers.

Acknowledgements The financial assistance for the work by the Council of Scientific and Industrial Research (CSIR), India, via sanction no. 03(1411)/17/EMR-II is thankfully acknowledged.

References

1. K. Takada, J. Power Sources **394**, 74 (2018)
2. H. Pan, S. Zhang, J. Chen, M. Gao, Y. Liu, T. Zhu, Y. Jiang, Mol. Syst. Des. Eng. **3**, 748 (2018)
3. F. Zheng, M. Kotobuki, S. Song, M.O. Lai, L. Lu, J. Power Sources **389**, 198 (2018)
4. X. Yao, B. Huang, J. Yin, G. Peng, Z. Huang, C. Gao, D. Liu, X. Xu, Chin. Phys. B **25**, 018802 (2016)
5. E. Zanotto, Am. Ceram. Soc. Bull. **89**, 19 (2010)
6. R.M. Mohamed, F.A. Harraz, I.A. Mkhalid, J. Alloy. Compd. **532**, 55 (2012)
7. K.B. Devi, B. Lee, A. Roy, P.N. Kumta, M. Roy, Mater. Lett. **207**, 100 (2017)
8. S. Brawer, *Relaxation in Viscous Liquids and Glasses* (American Ceramic Society, Columbus, OH, 1985), p. 38
9. K.S. Rao, P. Murali-Krishna, D.M. Prasad, L. Joon-Hyung, K. Jin-Soo, J. Alloy. Compd. **464**(1–2), 497 (2008)
10. N.F. Mott, E.A. Davis, *Electronic Processes in Non-Crystalline Materials* (Clarendon Press, Oxford, 1979)
11. H.W. Gibsen, R.J. Weagley, W.M. Prest, R. Mosher Jr, S. Kaplan, J. De Phys. Colloq. C3 **6**, 123 (1983)
12. J.C. Giuntini, J.V. Zancheha, J. Non-Cryst, Solids **34**, 419 (1979)
13. R. Vaish, K.B.R. Varma, J. Appl. Phys. **106**, 064106 (2009)
14. C.T. Moynihan, J. Non-Cryst. Solids **203**, 359 (1996)
15. R. Kundu, D. Roy, S. Bhattacharya, Phys. B **507**, 107 (2017)
16. P.W. Jaschin, K.B.R. Varma, J. Non-Cryst, Solids **434**, 41 (2016)
17. S. Singhal, J. Kaur, T. Namgyal, R. Sharma, Phys. B **407**, 1223 (2012)
18. S. Bhattacharya, A. Ghosh, J. Am. Ceram. Soc. **91**, 753 (2008)
19. M.L.-F. Nascimento, S. Watanabe, Braz. J. Phys. **36**, 795 (2006)
20. R. Bergman, J. Appl. Phys. **88**(3), 1356 (2000)
21. G.M. Parthun, G.P. Johari, J. Chem. Soc. Faraday Trans. **91**(2), 329 (1995)
22. M. Dult, R.S. Kundu, J. Hooda, S. Murugavel, R. Punia, N. Kishore, J. Non-Cryst. Solids **423–424**, 1 (2015)

Chapter 9
Optical Properties of Some Li-Doped Glassy Systems

Asmita Poddar, Madhab Roy, and Sanjib Bhattacharya

Abstract Glasses are one of the most important optical materials created to be transparent in the visible region. Most kinds of glassy system are prepared by the mixture of formers or intermediate oxides with other modifier oxides. Glasses containing rare earth ions are a subject of attention due to their many applications as laser materials, energy concentrators and luminescent materials. Different contents from lithium-doped glasses are prepared using the melt-quenching technique. The physical and structural properties of Li-doped glasses have been investigated in our work. Variations in the different physical parameters such as the density, molar volume, optical band gap, refractive index have been analysed and discussed in terms of the changes in the glass structure. The UV absorption spectra have been recorded at room temperature, and characteristics of optical energy band gap and Urbach Energy were determined for different glass systems.

9.1 Introduction

Glassy systems play an important role in the field of science, technology and industry [1]. In recent years, the development of glass–ceramics has extended the range of glass-based engineering materials [1]. They are important optical materials usually made to be transparent in the visible spectrum [1]. The structural, physical and optical characteristics of various glassy systems are greatly influenced by the composition and synthesis conditions [2]. Therefore, to accomplish high emission efficiency, most of the glassy system is activated using suitable transitional metals and/or rare earth

A. Poddar
Department of Electrical Engineering, Dream Institute of Technology, Kolkata 700104, West Bengal, India

A. Poddar · M. Roy
Department of Electrical Engineering, Jadavpur University, Kolkata 700032, West Bengal, India

S. Bhattacharya (✉)
UGC-HRDC (Physics), University of North Bengal, Darjeeling 734013, West Bengal, India
e-mail: sanjib_ssp@yahoo.co.in; ddirhrdc@nbu.ac.in

S. Bhattacharya and K. Bhattacharya (eds.), *Lithium Ion Glassy Electrolytes*,
https://doi.org/10.1007/978-981-19-3269-4_9

elements [2]. The incorporation of rare earth ions to the different glassy systems led to an improvement in the optical properties, such as refractive index, optical band gaps energy and laser amplification [2]. These improvements in optical properties for glassy systems drove them to be a potential candidate for lasers, solar concentrate systems, optical detectors and waveguides and telecommunications optical fibres. Glasses doped with rare earth ions have been investigated intensively for photonics, optoelectronics and scintillating applications [2]. Borate glasses have specific properties like coordination geometry, high transparency, higher bond strength, low melting point and good rare earth ions solubility that make them beneficial for a wide technical application [2]. But, their chemical durability is relatively feeble, which limits their utility. It has been studied that the addition of oxides such as lithium oxide (Li_2O) and aluminium oxide (Al_2O_3) can enhance the chemical durability and physical properties [3]. Moreover, adding metal oxides as modifiers to the host matrix raises the radiative parameters. Besides, glasses containing metals minimize phonon energy and lead to an increase in the luminescence quantum from excited rare earth ion states. Therefore, glasses doped with rare earth elements may be used because of their ion emission efficiencies. Due to their unique characteristics, glass systems doped with rare earth elements have been given considerable attention in recent decades. Luminescence, lasing and sensing properties of rare earth (RE) ions have drawn abundant interest among the researchers [3]. Rare earth-doped glasses have potential applications due to their emission efficiencies of electronic transitions in the RE ion [3]. These glasses are marvellous luminescent materials owing to the occurrence of sharp fluorescence in ultraviolet (UV), visible and infrared (IR) regions due to their shielding effects of the outer electron. Rare earth-doped glasses are potential candidates for laser hosts, waveguide, optical fibres, solar concentrators, plasma display panel, optical amplifiers, semiconductor light-emitting diodes, optical detectors and so forth.

9.2 Experimental Results and Analysis

Glasses obtained from a variety of nonlinear materials are widely used for multi-facets applications in modern technology due to their unusual physical and optical properties. The investigation of the changes in the physical properties of glasses with controlled variation of chemical composition and dopants of transition metal ions is of considerable interest in the application point of view. Studies have shown that physical properties of the glass can be improved with the addition of Li, which increases humidity resistance and capacity to concentrate transition metal ions [4].

Following are the experimental results that have obtained when researches on some Li-doped glassy systems have been carried out:

9.2.1 *Studies on Density, Molar Volume, Coordination Number and Refractive Index*

It has been found that the density increases remarkably with addition of Li_2O in the borate glass system [1]. The density is affected by the structural softening or compactness, change in coordination number, cross-link density and dimension of interstitial spaces of glass. In the lithium borate glass system $(Li_2O)_x\ (B_2O_3)_{1-x}$, the tetrahedral BO_4 groups that form at low lithium content are stronger bonded compared to the triangular BO_3 groups and increase its density. Lithium oxide acts as modifier in the glassy network. When Li_2O is added into glass network, transition from tetrahedral BO_4 groups to triangular BO_3 groups with non-bridging oxygen (NBO) occurs. The increase in density is due to increase in the number of non-bridging oxygen (NBO) atoms.

Molar volume decreases with content of Li_2O, which is shown in Fig. 9.1, is because of the compact structure as well as due to the fact that Li_2O occupies interstitial position in the network.

In this ground, a study of refractive index of lithium borate glassy system has been carried out, and it has been found that the index increases with Li_2O as shown in Fig. 9.2. It can be clearly seen that refractive index is inversely proportional to the molar volume. Thus, the refractive index increases with decreasing molar volume and which in turn increases the density.

Again, the coordination number of lithium borate glass also leads to the increment in refractive index [20]. An addition of Li_2O causes change in coordination number and creates more non-bridging oxygen. Thus, a higher average coordination

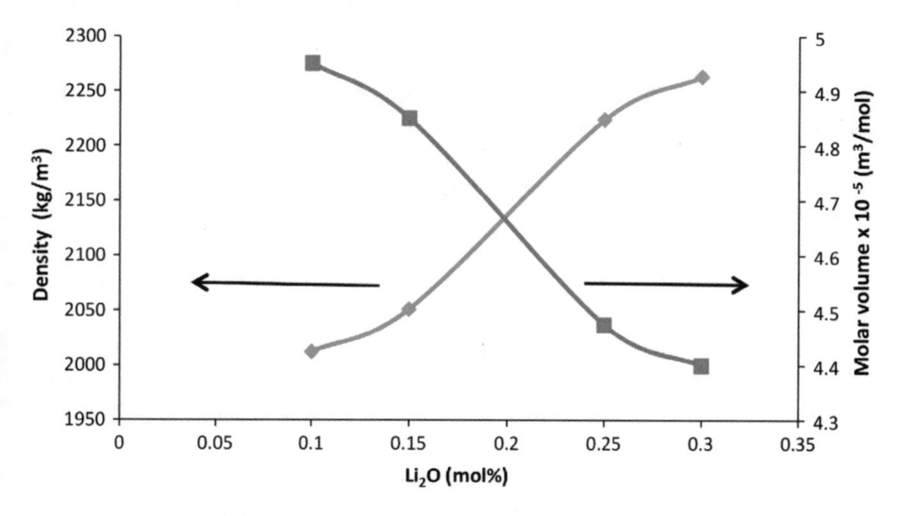

Fig. 9.1 Density and molar volume of $(Li_2O)_x-(B_2O_3)_{1-x}$ glassy system. Republished with permission from Halimah et al. [1]

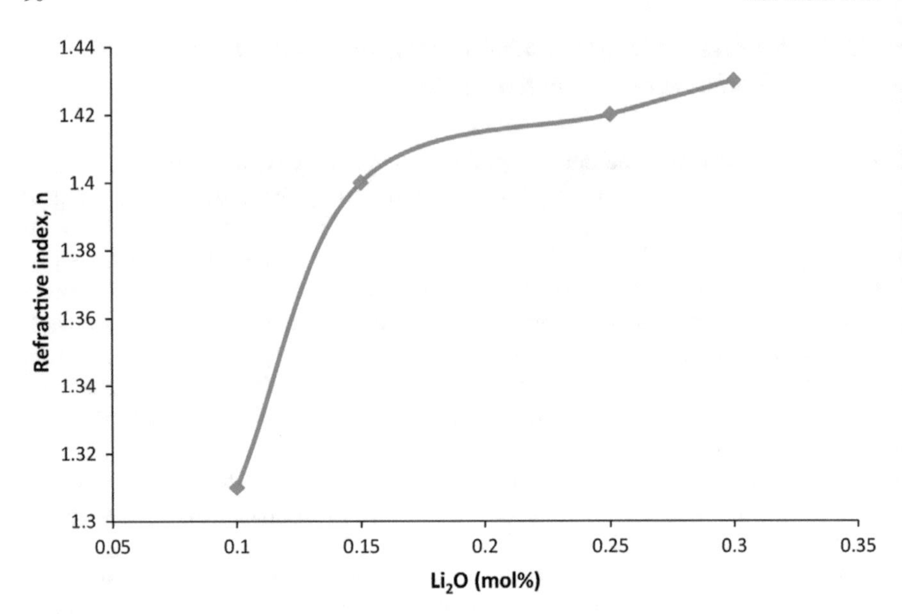

Fig. 9.2 Refractive index $(Li_2O)_x-(B_2O_3)_{1-x}$ glassy system. Republished with permission from Halimah et al. [1]

number of studied glass may exhibit higher values of refractive index. The formation of non-bridging oxygen forms more ionic bonds, which manifest themselves in a large polarizability, thus results in a higher index value [1]. But, another study has been done on bismuth–silicate glasses containing lithium oxide, and this study shows some different results [5]. A glassy system having composition $xLi_2O{\cdot}(85-x)Bi_2O_3{\cdot}15SiO_2$ $(5 \leq x \leq 45$ mol%) has been prepared, and density, molar volume and glass transition temperature for the samples are measured. The study on variations of density as well as the molar volume of all the samples is shown in Fig. 9.3. It can be seen clearly that the density values decreases with the increase in Li_2O content. These data are expected because of the high mass of the heavy metal cation (Bi^{3+}) [5]

Studies have been done on a glass system containing Cr^{3+}-doped 19.9 ZnO + $xLi_2O + (30-x)\ Na_2O + 50B_2O_3$ $(5 \leq x \leq 25)$ [4]. With an increase in Li_2O content, it can be seen that the above physical parameters vary prominently.

Figure 9.4 shows the compositional dependence of density and refractive index of Cr^{3+}-doped ZLNB glassy systems [4]. The density values reach a maximum value at $x = 10$ mol% from a minimum value at $x = 5$ mol% and then decrease and increase with increase of x mol%.

Similarly, the refractive index values also decrease and increase with increase of x mol%, and the maximum value of refractive index is observed at $x = 15$ mol%.

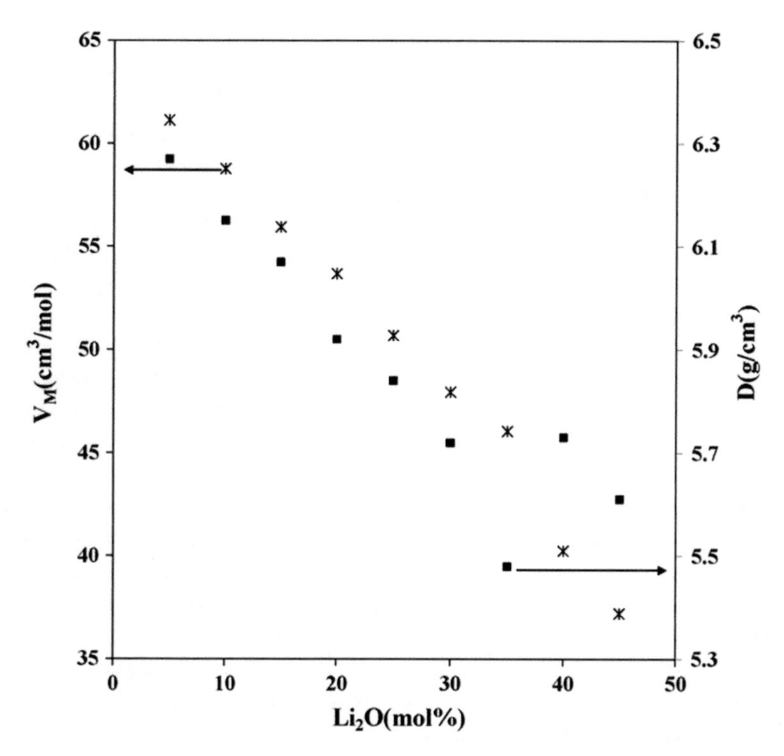

Fig. 9.3 Composition dependence of density and molar volume of lithium bismosilicate glasses. Republished with permission from Ahlawat et al. [5]

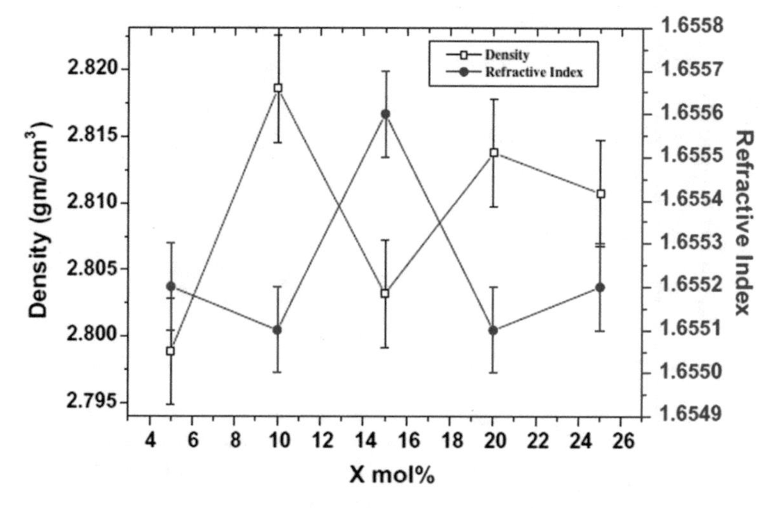

Fig. 9.4 Correlation between density and refractive index of Cr^{3+}-doped ZLNB glasses. Republished with permission from Rama Sundari et al. [4]

9.2.2 Study on Optical Band Gap

For understanding and developing the band structure and energy band gap (E_g) of crystalline and non-crystalline material structures, the analysis of optical absorption spectrum has been proved as one of the most productive tools [4]. There are two types of optical transitions that can occur at the fundamental absorption edge of crystalline and non-crystalline materials [4]. They are direct and indirect transitions. In the case of glasses, the conduction band is influenced by the anions, and the cations play an indirect but major role.

Study on optical band gap of lithium borate glasses shows that the direct band gap values are larger than the indirect band gap, and both values decrease with the increase of Li_2O content [1]. Figure 9.5 shows the optical band gap values for $(Li_2O)_x(B_2O_3)_{1-x}$ system vary between 0.1 and 2.3 eV for indirect transition and vary from 3.8 to 4.8 eV for direct band gap [1].

To study the nature of lithium borate glasses in more distinct way, chromium has been doped in it [4], and the optical band gap study has been done. From the band gap energy calculations, it can be inferred that the glass systems are direct semiconductors. The investigations show that absorption edge is more in this system when it is compared to divalent ions-doped ZLNB glasses [6–8]. This may be due to incorporation of Cr^{3+} in to the host glass network. There are studies which show chromium-doped ZLNB glasses and show more absorption edge and optical band gap energies similar to other glass systems [9–11] too.

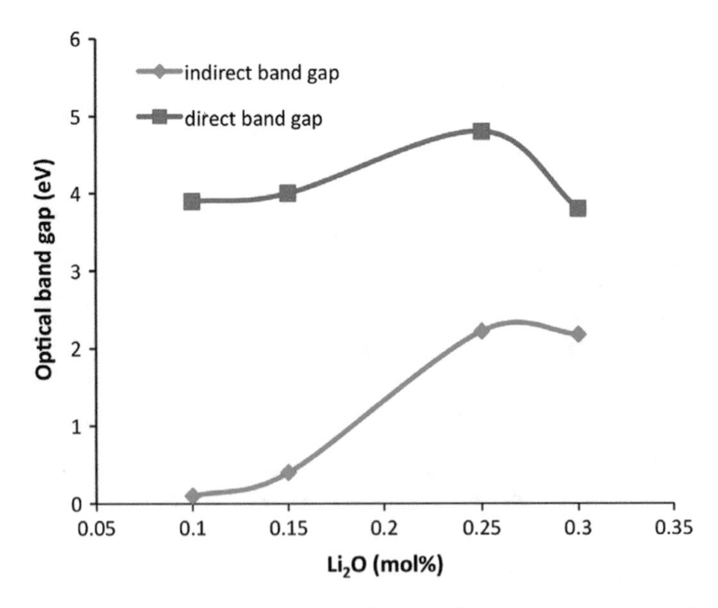

Fig. 9.5 Indirect and direct optical band gap of $(Li_2O)_x-(B_2O_3)_{1-x}$ glassy system. Republished with permission from Halimah et al. [1]

The effect of optical band gap energy on bismuth–silicate glasses containing lithium oxide has been also studied [5]. And the studies show that with the increase of Li_2O content, the optical band gap energy E_{opt} first increases and then decreases [5]. The increase in lithium ions causes structural change in the glass network [5]. The formation of non-bridging oxygen (NBO), which binds excited electrons less tightly than bridging oxygen, increases with Li_2O which results in the decrease of optical energy band gap [5]

The optical transmission spectra of the samples are generally recorded at room temperature, and the effect of lithium oxide on the optical absorption edge can be evaluated. The effects that can be seen in bismosilicate glasses [5] are shown in Fig. 9.6. It is observed that optical absorption edge is not sharply defined, which concludes that the samples are amorphous in nature. It is also observed that the cut-off wavelength shifts towards longer wavelength as the content of Li_2O increases beyond 35 mol%. This shift may be attributed to the increase in number of the non-bridging oxygen ions beyond a particular ratio of Bi_2O_3/Li_2O [5].

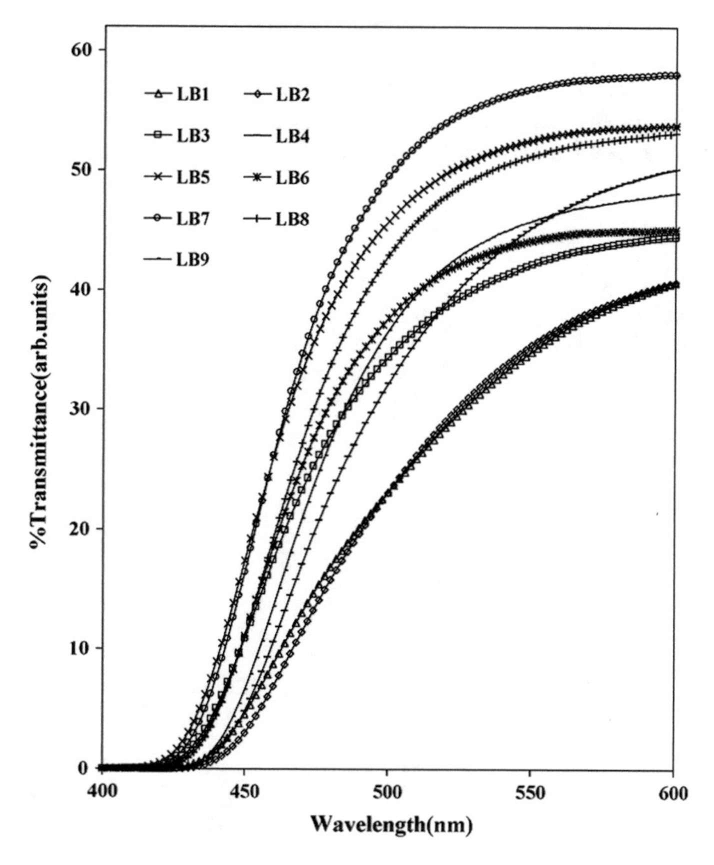

Fig. 9.6 Optical absorption spectra for $x Li_2O \cdot (85 - x) Bi_2O_3 \cdot 15 SiO_2$ glasses. Republished with permission from Ahlawat et al. [5]

The optical band gap study shows that although the conduction band is mainly influenced by the glass-forming anions, the cations also play a significant role indirectly. The variations in optical band gap (E_g) values with the alkali content may be because of indirect influence of Li_2O on the band gap [12].

Study of optical band gap is also carried out on lithium halo borate glasses, prepared from appropriate mixtures of H_3BO_3, BaF_2, LiF, LiCl and LiBr, and it shows some interesting features [13]. When the relation between E_{opt} and LiX concentration of the samples is studied, it clearly shows that there is sharp decrease in E_{opt} in the range from 0 to 5mol% LiX as shown in Fig. 9.7. At concentrations more than 5mol%, the decreasing rate is much smaller than in the range 0–5 mol%. The reason of the sharp decrease in the range 0–5 mol% LiX may validate the fact that the halide ions substitutes oxygen so as to create more weaker BO_4 and/or oxyhalide groups like BO_2F, BO_2F_2, BOF_3 and BO_3F [14]. At LiX concentrations more than 5 mol%, halide ions occupy mainly interstitial positions, and the structural changes are gentler. The decrease in E_{opt} with LiX content can be understood in terms of the structural changes taking place in the glasses. According to research [15] when the concentration of LiX is zero, the number of borate rings is greater than the number of boroxol rings. In the pure borate group, there is boroxol ring oxygen breathing vibration involving a very little boron motion. When alkali ion is present in a small amount, there is an increase in the number of BO_4 units against the number of BO_3 units. With further addition of LiX, the six-membered borate rings with only one BO_4 tetrahedron appear.

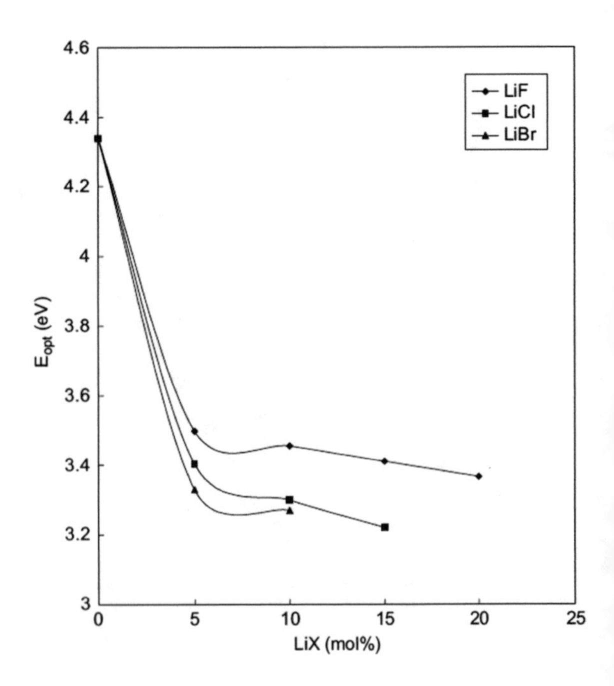

Fig. 9.7 Variation of optical energy gap E_{opt} with composition of the studied glasses. Republished with permission from Hager [13]

9.2.3 UV Absorption Spectra

The lithium halo borate glasses have also undergone study of UV absorption spectra [13], and the results show that increasing LiX concentration shifts the UV absorption edge to lower energies (higher wavelengths) as shown in Fig. 9.8a, b for different systems. This shift occurs because of increment in the concentration of NBO ions with increasing LiX content in the glassy system.

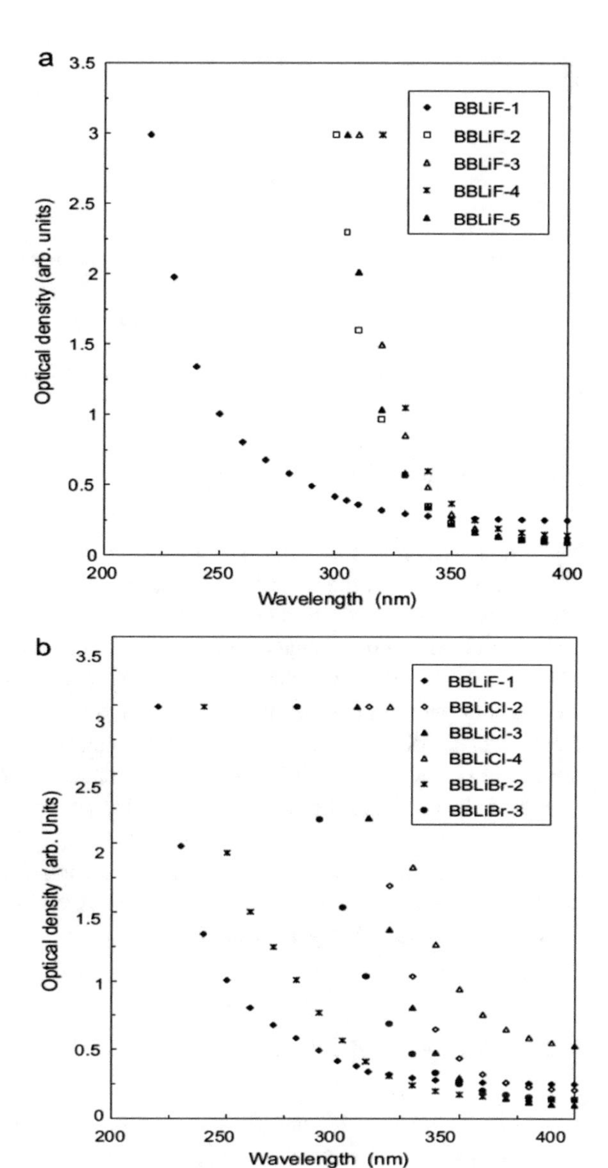

Fig. 9.8 **a** UV absorption spectra of B_2O_3–BaF_2–LiF glass and **b** UV absorption spectra of B_2O_3–BaF_2–LiX (X = Cl and Br) glasses. Republished with permission from Hager [14]

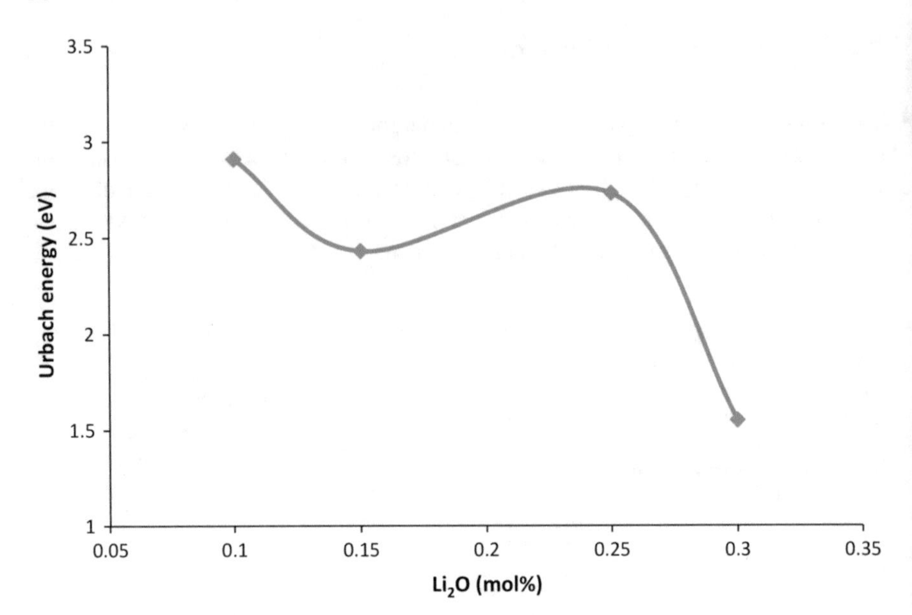

Fig. 9.9 Urbach Energy of $(Li_2O)_x-(B_2O_3)_{1-x}$ glassy system. Republished with permission from Halimah et al. [1]

9.2.4 Study of Urbach Energy

The Urbach Energy is an important factor when to deal with glass systems [1]. It gives information on the disorder effects, and the research work shows that it also changes with Li_2O content in the glassy system [1].

Urbach Energy corresponds to the width of localized states and is used to characterize the degree of disorder in amorphous and crystalline systems. If the energy of the incident photon is less than the band gap, there is increase in absorption coefficient which is followed with an exponential decay of density of localized states into the gap [25], and the edge is known as the Urbach Energy.

Materials with larger Urbach Energy would have greater tendency to convert weak bonds into defects [21]. Figure 9.9 shows that with increase of Li_2O content, the Urbach Energy starts decreasing, and this decreasing trend suggests that the degree of disorder gets also decreased. Smaller is the value of Urbach Energy, greater is the structural stability of the glass system.

9.2.5 Study of Infrared Spectra

Infrared spectroscopy is one of the most useful experimental techniques available for easy structural studies of glasses [16]. This technique leads to structural aspects related to both the local units constituting the glass network and the anionic sites hosting the modifying metal cations.

Infrared is a powerful tool for the structural studies of glasses modified by metal oxides. Thus, when the above-discussed Li-doped glass systems have undergone infrared transmission spectra study, it has revealed the most useful data of the prepared glass systems.

In the study [5], the mid- and near-infrared spectra of the bismuth–silicate glasses containing lithium oxide glass system show some resemblance with the spectra usually obtained from the traditional silicate glasses and crystals [17]. But the positions of absorptions bands are different due to the abundance of the heavy metal bismuth oxide (Bi_2O_3) and modifying cation (Li_2O). From the IR spectra study, it can be inferred that the broadband shifts towards longer wave number as bismuth decreases and Li_2O content increases which suggests loosening of the glass network leading to decrease in glass transition temperature. This shifting with increment of Li_2O content is related to the change of local symmetry, and the theory is accepted by various authors [18, 19]. It infers that the band shifting to higher wave number (471 cm^{-1}) is due to the increase of the degree of distortion. When the Fourier transform infrared (FT-IR) study has been done on Cr^{3+}-doped ZLNB glass system [4], it shows interesting features. The IR analysis exhibits four different structural information as shown in Fig. 9.10.

The obtained absorption bands and their assignments can be summarized as:

(i) In the region 600–800 cm^{-1}, the bands are due to the bending vibrations of B–O–B linkages.
(ii) In the region 800–1140 cm^{-1}, the bands are identified due to the B–O symmetric stretching vibrations of BO_4 units.
(iii) Bands observed in the region 1200–1520 cm^{-1} are assigned to asymmetric stretching vibrations of B–O units in different borate groups.
(iv) The vibrational band observed at around 1679 cm^{-1} is ascribed to the vibrational modes of hydroxyl or water groups present in the glass systems.

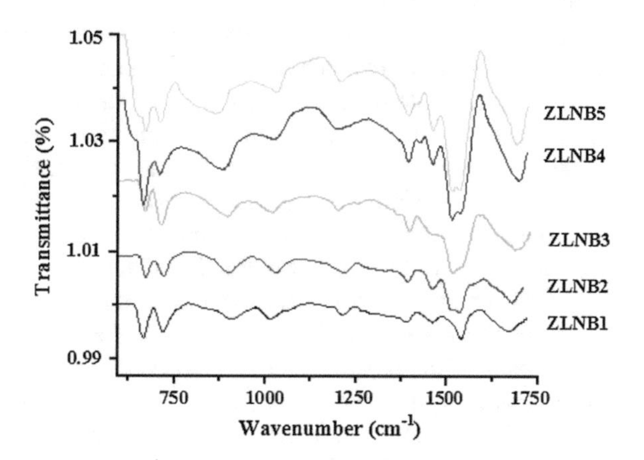

Fig. 9.10 FT-IR spectra of Cr^{3+}-doped ZLNB glassy systems. Republished with permission from Rama Sundari et al. [4]

9.3 Review Works and Applications

Lithium halo borate glasses have been prepared according to the formula (70 – y) B_2O_3–30BaF_2–yLiX where $y = 0, 5, 10, 15$ and 20 mol% and X = F, Cl and Br [13]. The study shows that edge of the UV absorption spectra of lithium halo borate glass shifts to lower energy with increasing LiX concentration. E_{opt} gets decreased with increment of B–O bond length and the radius of the halogen ion. Increasing the number of interstitial halide ions and the creation of orthoborate groups raises the glass disorder and the Urbach Energy E_{tail}. The optical energy band gap decreases with increase in LiX content due to the increase in NBO concentration.

Bi_2O_3-based glasses are popular because of their wide applications in the field of glass ceramics, layers for optical and optoelectronic devices, thermal and mechanical sensors, reflecting windows, etc., [20]. When the studies have been done on Li_2O-doped borobismuthate glasses having composition 25Li_2O–(75–x) Bi_2O_3–x B_2O_3, it shows that bismuthate glasses containing alkali oxide act as ionic conductors and possess high conductivity compared to other heavy metal glasses [20].

The density of the glass decreases due to the lower atomic weight of B_2O_3. The average electronic oxide ion polarizability and optical basicity of the studied glasses have been estimated on the basis of refractive index and optical band gap [20]. The IR analysis of these glasses confirmed the decrease in the molar volume, which in turn explains how the refractive indices, polarizability and the optical basicity of them decrease with compositions.

The optical absorption spectra of the glasses having composition $(25+x)Li_2O{\cdot}(65-x)P_2O_5{\cdot}10Bi_2O_3$ ($0 \le x \le 25$ mol%) were recorded in spectral range from 200 to 3300 nm at room temperature [21]. The optical absorption edge, optical band gap and Urbach Energy were determined from the absorption spectra, and it has been shown that variation in these optical parameters has been associated with the structural changes occurring in these glasses with increase in Li_2O: P_2O_5 ratio.

The study of IR spectra of such glassy systems [21] has revealed that the optical band gap of the sample decreases with increase in Li_2O and reaches up to semiconductors' range. The increase in density and decrease in molar volume with increase in Li_2O:P_2O_5 ratio have been related to the changes in the glass structure [21]. Long chains of phosphate groups are disrupted by the introduction of Li_2O into the glass matrix resulting in the lowering of glass transition temperature [21].

Study of Raman and IR spectra [5] has been done to investigate the structure of unconventional lithium bismuthate glasses over the wide range of alkali oxide and found that the structure of this unconventional binary glass system changes systematically with the increase of Li_2O content.

SiO_2 is one of the most common glass formers, and its glass-forming range can be extended by addition of alkali oxide [5]. The oxygen in the Si–O–Si linkage is known as bridging oxygen (BO), and oxygen in Si–O – is known as non-bridging oxygen (NBO). The alkali ions (Li_2O) locate themselves in the structure near the NBO's [5]. The degradation of network is assumed to be systematic as the alkali concentration increases [5].

Lithium–cesium borate glasses doped with chromium ions have been extensively studied by a group of workers [22]. This study [22] reveals that the optical band gap energy increases with compositions and reaches a minimum around $x = 15$ mol%, and thereafter, it increases. It is observed that the optical band gap energies vary from 2.20 to 2.76 eV for both the direct and indirect transitions [22]. It is evident that Urbach Energy (0.29 eV) is minimum for $x = 5$ mol% in the glass samples [22].

Mixed alkali bismuth borate glasses $xLi_2O–(30 - x) K_2O–10Bi_2O_3–55 B_2O_3$ ($0 < x < 30$) doped with 5 mol% vanadium ions were prepared from the melts [23]. Optical absorption studies were carried out as a function of alkali content to look for mixed alkali effect on the spectral properties of these glasses. From the study of ultraviolet absorption edge, the optical band gap energies and Urbach Energies were evaluated. The average electronic polarizability of the oxide ion, optical basicity and the interaction parameters were also evaluated for all the glasses. Many of these parameters vary nonlinearly exhibiting a minima or maxima with increasing alkali concentration, indicating the mixed alkali effect. An attempt is made to interpret mixed alkali effect in this glass system in terms of its glass structure.

Researchers have shown that glasses containing Cr_2O_3 have interesting optical properties due to the presence of chromium ions in two possible oxidation states: trivalent and hexavalent forms. So, studies on Cr^{3+}-doped $19.9 ZnO + xLi_2O + (30 - x) Na_2O + 50B_2O_3$ ($5 \leq x \leq 25$) glassy system [4] have been done, and it reveals that FT-IR spectral analysis confirms the presence of BO_3 and BO_4 local structures in the concerned system. The optical absorption spectra confirm the distorted octahedral site symmetry for Cr^{3+} ions with partial covalency of ZLNB glasses. From optical absorption edges, optical band gap energies and Urbach Energies were evaluated, and the nonlinear behaviour is observed, which is in good agreement with theoretical values [4]. It also confirms the structural variations with x mol% of alkali ions.

A study on structural and optical properties of lithium tetraborate glass has also been done [24]. The samples containing chromium and neodymium having formula $99 Li_2B_4O_7 - (1 - x) Cr_2O_3 -x Nd_2O_3$ mol% (where $x = 0, 0.25, 0.50, 0.75$ and 1 mol %) were prepared, and density, molar volume, oxygen packing density have been estimated for the system [24]. Neodymium is one of the important elements that has been used in a variety of host glasses in which laser action has been observed [24]. Studies have shown that these components are able to form a transparent glass suitable for optical applications. The FT-IR absorption spectra of the prepared samples [24] measured at room temperature show that the value of E_g decreases as the Nd content increases up to 0.75 mol% and then increases. The decrease of E_g can be related to the change of bridging oxygen to non-bridging oxygen, and these results agree with density and molar volume of the system. As the concentration of octahedral Cr^{3+} ions increases in the system, the concentration of non-bridging oxygen (NBO) also increases in the glassy matrix. This leads to an increase in the degree of localization of electrons and thus decreasing the donor centres in the glass matrix. The presence of larger concentration of these donor centres increases the optical band tail and shifts the absorption edge towards low wavelength.

9.4 Conclusion

Rare earth ions doped with different host glasses are used for various optical devices like colour displays, lasers, fibre amplifiers, solid-state lighting devices, etc. in the past years, and researches are going on to find out more dominant characteristics of these glass systems in the field of optics. The physical and optical properties of such glassy sample were dependent on the glass composition. From the studies, it is evident that the increment in density of Li_2O content has great effects on the number of non-bridging oxygen. The molar volume and density of the glasses change since Li_2O occupies interstitial position in the network. The formation of non-bridging oxygens results in the decreased value of optical energy band gap for both direct and indirect band gaps. This is because non-bridging oxygen binds excited electrons less tightly than bridging oxygen. Here, we have reported the physical parameters, FT-IR, Raman spectra, optical energy band gap, refractive index, optical basicity, electronic polarizability and luminescence properties related to various compositions of Li-doped glasses. Significant efforts have been made to explore the microstructure, glass-forming range and stability, humidity withstanding ability associated with chemical durability and transparency from UV to NIR range of Li_2O-doped various glassy systems not only for their applications point of view but also for academic interest.

References

1. M.K. Halimah, W.H. Chiew, H.A.A. Sidek, Optical properties of lithium borate glass $(Li_2O)x$ $(B2O3)1-x$, Sains Malaysiana **43**(6), 899–902 (2014)
2. Bagi Aljewaw O, Karim MK, Mohamed Kamari H, Mohd Zaid MH, Mohd Noor N, Che Isa IN, Abu Mhareb MH, Impact of Dy_2O_3 substitution on the physical, structural and optical properties of lithium–aluminium–borate glass system. Appl Sci **10**, 8183 (2020)
3. P.P. Pawar, S.R. Munishwar, S. Gautam, R.S. Gedam, Hysical, thermal, structuralandopticalpropertiesofDy3þ doped lithium alumino-borate glasses for bright W-LED. J. Lumin. **183**, 79–88 (2017). https://doi.org/10.1016/j.jlumin.2016.11.027
4. G. Rama Sundari, V. PushpaManjari, T. Raghavendra Rao, D.V. Satish, Ch.Rama Krishna, C.H. Venkata Reddy, R.V.S.S.N. Ravikumar, Optical Mater. **36**, 1329 (2014)
5. N. Ahlawat, S. Sanghi, A. Agarwal, S. Rani, Effect of Li_2O on structure and optical properties of lithium bismosilicate glasses. J. Alloys Compd **480**, 516–520 (2009)
6. T. Raghavendra Rao, C.H. Rama Krishna, C.H. Venkata Reddy, U.S. Udayachandran Thampy, Y.P. Reddy, P.S. Rao, R.V.S.S.N. Ravikumar, Spectrochimica. Acta A **79**, 1116–1122 (2011)
7. T. Raghavendra Rao, C.H. Venkata Reddy, C.H. Rama Krishna, U.S. Udayachandran Thampy, R. Ramesh Raju, P. Sambasiva Rao, R.V.S.S.N. Ravikumar, J. Non-Cryst. Solids **357**, 3373–3380 (2011)
8. T. Raghavendra Rao, C.H. Venkata Reddy, C.H. Rama Krishna, D.V. Sathish, P. Sambasiva Rao, R.V.S.S.N. Ravikumar, Mater. Res. Bull. **46**, 2222–2229 (2011)
9. G. Naga Raju, N. Veeraiah, G. Nagarjuna, P.V.V. Satyanarayana, Physica B **373**, 297–305 (2006)
10. S. BalaMurali Krishna, P.M. Vinaya Teja, D. Krishna Rao, Mater. Res. Bull. **45**, 1783–1791 (2010)
11. G. Venkateswara Rao, N. Veeraiah, J. Alloys Compd. **339**, 54–64 (2002)

12. M. Vithal, P. Nachimuthu, T. Banu, R. Jagannathan, J. Appl. Phys. **81**, 7922 (1997)
13. I.Z. Hager, Optical properties of lithium barium haloborate glasses. J. Phys. Chem. Solids **70**(1), 210–217 (2009)
14. I.Z. Hager, M. El-Hofy, Phys. Status Solidi. (a) **198**(1), 7 (2003)
15. D. Maniu, I. Ardenlean, T. Iliescu, Mater. Lett. **25**, 147 (1995)
16. J. Wong, C.A. Angell, *Glass Structure by Spectroscopy* (Marcel Dekker Inc., New York, 1967), p. 409
17. A. Witkowska, J. Regbicki, A.D. Eicco, J. Alloys Compd. **324**, 109 (2003)
18. L. Baia, R. Stefan, W. Kiefer, J. Poppe, S. Simon, J. Non-Cryst. Solids **303**, 379 (2002)
19. F.H. El Batal, Nucl. Instrum. Methods Phys. Res. B **254**, 243 (2007)
20. E.S. Moustafa, Y.B. Saddeek, E.R. Shaaban, Structural and optical properties of lithium borobismuthate glass. J. Phys. Chem. Solids **69**(9), 2281–2287 (2008)
21. S. Rani, S. Sanghi, A. Argawal, V.P. Seth, Study of optical band gap and FTIR spectroscopy of Li2O.Bi2O3.P2O5 glasses. J. Non Cryst. Solid **74**(3), 673–677 (2009)
22. C.R. Kesavulu, R.P.S. Chakradhar, C.K. Jayasankar, R.J. Lakshmana, ERP, optical, photoluminescence studies of Cr^{3+} ion in Li_2O-Cs_2O-B_2O_3 glasses—An evidence of mixed alkali effect. J. Mol. Struct. **975**(1–3), 93–99 (2010)
23. M. Subhadra, P. Kistaiah, Characterisation and optical absorption studies of VO^{2+}: Li_2O-K_2O-Bi_2O_3-B_2O_3 glass system. J. Alloy Compd. **505**, 634–639 (2010)
24. I. Kashif, A. Ratep, Saffa K. El-Mahy, Structural and optical properties of lithium tetraborate glasses containing chromium and neodymium oxide. https://doi.org/10.1016/j.materresbull.2017.02.006
25. B. Abay, H.S. Gudur, Y.K. Yogurtchu, Solid State Commun. **112**, 489 (1999)

Chapter 10
Mechanical Properties of Some Li-Doped Glassy Systems

Ajit Mondal, Debasish Roy, Arun Kumar Bar, and Sanjib Bhattacharya

Abstract The mechanical properties of lithium-doped glassy materials depend on various factors. We have also discussed the different types of lithium-doped materials such as $Li_2O–Al_2O_3–SiO_2$-based glass–ceramic, lithium aluminum silicate glass–ceramics, $Li_2O–ZnO–SiO_2$ glass–ceramics, $Li_2O–Na_2O–K_2O–ZnO–B_2O_3$, lithium magnesium borate glass. The chemical composition, crystallinity, density (ρ), molar volume (Vm), residual stress, additives, nucleating agents, crystal size, elastic properties and microstructural analysis significantly play important role on the mechanical properties of lithium-doped glassy materials.

Keywords Lithium-doped glassy materials · Mechanical properties · Lithium disilicates

10.1 Introduction

In recent years, some glassy materials are being applied in numerous industrial fields, such as bioactive implants [1], low thermal expansion materials [2] and dental applications [3]. It is also utilized in various types of glass–ceramics like silicates, alumino silicates and fluosilicates grouped by compositions [2]. Lithium disilicate (LS_2) crystal-based glass–ceramics are commercially successful so far in dental applications due to their excellent mechanical properties [1–3]. Lithium disilicate was firstly classified as glass–ceramic material by Stookey in the year 1959 [4].

A. Mondal
Department of Automobile Engineering, Raiganj Polytechnic, Uttar Dinajpur, West Bengal 733134, India

D. Roy
Department of Mechanical Engineering, Jadavpur University, Kolkata, West Bengal 700032, India

A. K. Bar
Department of Mechanical Engineering, Institute of Engineering and Management, Kolkata, India

S. Bhattacharya (✉)
UGC-HRDC (Physics), University of North Bengal, Darjeeling, West Bengal 734013, India
e-mail: ddirhrdc@nbu.ac.in

S. Bhattacharya and K. Bhattacharya (eds.), *Lithium Ion Glassy Electrolytes*,
https://doi.org/10.1007/978-981-19-3269-4_10

It was developed by Beall in the year 1971 [5] and Freiman and Henchin in the year 1972 in multicomponent glass systems [6]. Structure of lithium disilicate glass and its corresponding crystal was well defined by the Hannon et al. in the year of 1992 [7] and Soares Jr et al. in the year of 2003 [8]. Lithium disilicate glass–ceramics showed some the properties. Among them, mentioning few properties that are flexural strength varies between 300.00 MPa and 400.00 MPa, and the next property is fracture toughness that is between 2.80 MPa. $\sqrt{m}$ and 3.50 MPa. $\sqrt{m}$. That is why they are currently used in the following fields [9–12]:

(i) In the field of dental restorations
(ii) In the field of space maintainers
(iii) Application in replacement of tooth appliances
(iv) Splints
(v) Dental crowns
(vi) Partial crowns
(vii) To make dentures
(viii) Posts
(ix) To make inlays and on lays
(x) To make laminate veneers
(xi) In the application of implant abutments and the domain of restorations.

10.2 General Consideration

Lithium disilicates have good press ability. Injection molding helps it to be into refractory investment molds. In this process, lost wax technique is utilized. Lithium aluminum silicate (LAS) glass–ceramics bear a range of many useful properties such as high strength, proper resistance to mechanical and thermal shocks and worthy chemical durability [13]. As a result, LAS glass–ceramics show extensive application in heat exchangers and cookware and also in the field of telescope mirror supports, etc. These high strength and high thermal expansion coefficient (TEC) glass–ceramics are utilized in hermetic sealing and also enameling with high TEC metals. To shortly tune the TEC and several thermo-mechanical properties, heat treatment schedule for crystallization must be seriously optimized [14–16]. Glass–ceramics have polycrystalline materials fabricated by limited crystallization of proper glasses according to a heat treatment process that facilitates nucleation and also crystal growth. The resulting materials are composed of one or various phases of crystalline impacted in a glassy matrix that firstly had enhanced wear resistance, secondly chemical resistance, thirdly flexural strength and hardness, fourthly fracture toughness and also including dimensional stability compared with non-ceramic glasses [1–6]. Microstructure, composition, crystallinity, thermal history, kinetics, phase composition and additives are the several factors which determine the physical properties of glass–ceramics. The properties of brittle materials especially the mechanical properties depend upon some factors. One of them is the internal micromechanical stresses. Actually, elastic and thermal mismatch that occurs amid the constituent phases is reason of this internal micromechanical stress. All these occurrences in brittle materials take place on cooling [17, 18]. The optical properties rely on the crystal size, the differences in the refractive index of the glass matrix and also crystalline phase. As crystal size decreases, translucency increases. The glass matrix has the similar refractive index as the crystalline phase which leads to translucency to increase. In contrast, large crystals decrease the translucency but at the same time increase the mechanical properties of glass–ceramics [19].

10.3 Few Lithium-Doped Glassy Systems

Bo et al. [20] developed the Li_2O–Al_2O_3–SiO_2-based glass–ceramic with a low softening point, which could be sintered below 900 °C. Arvind et al. [16] investigated the thermo-mechanical properties of lithium aluminum silicate glass–ceramics. The thermo-physical properties in Li_2O–ZnO–SiO_2 glass–ceramics were explored by Sharmaa et al. [21]. The effect of optical, structural, thermal and mechanical properties of Li_2O–Na_2O–K_2O–ZnO–B_2O_3-based glassy system had been showed by Subhashini et al. [22]. The impact of physical and optical properties of lithium magnesium borate glassy system had been examined by Mhareb et al. [23]. Salman et al. [24] explored the effect of Al_2O_3, MgO and ZnO on the crystallization characteristics and properties of lithium calcium silicate glasses and glass–ceramics.

10.4 Different Mechanical Properties of Some Lithium-Doped Glassy Systems (Literature Survey)

The chemical composition, crystallinity, additives, residual stress, nucleating agents, aspect ratio, crystal size and morphologies greatly affect its mechanical properties—this had been shown in a considerable number of studies [25–28]. For example, Denry et al. [29] reviewed that high crystallinity and great aspect ratio crystalline grains generated interlocking microstructures that facilitated crack bridging and deflection in several lithium disilicate glass–ceramics. Buchner et al. [30] showed that the densification and crystallization under high pressure and temperature enhance the mechanical properties of lithium disilicate glass–ceramics. Goharian et al. [31] had observed three-point flexural strength of 280 MPa for a lithium disilicate glass–ceramics sample. Hölland et al. [32] practiced the microstructure and properties of different dental glass–ceramics and indicated a moderate fracture toughness value for lithium disilicate glass–ceramics. The mechanical properties of lithium disilicate glass–ceramic had been explored by Sang-Chun [33] and indicated a fracture toughness and flexural strength values of about 3.2 MPa and about 334 MPa, respectively. The elastic constants, fracture toughness and Vickers hardness of lithium disilicate glass–ceramic had been investigated by Denry and Holloway [34].

10.4.1 Mechanical Testing

Li et al. [26] used the plate-shaped glass–ceramic specimens for three-point bending tests (guidelines of ISO 6872) [35]. The tests had done by the screw-driven SUNS CMT4204 testing machine at a constant cross-head displacement rate of 0.50 mm/min. The flexural strength (σ) had been determined by the following relation:

$$\sigma = \frac{3Pl}{2wb^2} \tag{10.1}$$

where

P	breaking load,
l	test span (15 mm),
w and b	width and thickness of the specimens, respectively.

The fracture toughness (K_{IC}) is defined as ability of a crack to propagate within the materials taking side by side the applied load. The fracture toughness (K_{IC}) could be calculated by a certain which was given by Anstiset al. [36]

$$K_{\mathrm{IC}} = \sqrt{\left(\frac{E}{H}\right)\left(\frac{P}{\sqrt{C^3}}\right)} \tag{10.2}$$

where

K_{IC}	is measured in MPa,
E	Young's modulus (GPa),
H	Vickers hardness (GPa) as determined by hardness,
P	applied load (4.90 N),
C	half crack length (μm) as measured.

Modulus of elasticity of the polycomponent system was determined with the help of model (theoretical) given by Makishima and Mackenzie [37]:

$$E = 83.60\, V_t \sum_i G_i X_i \tag{10.3}$$

where, G_i = dissociation energy density of oxide constituents, X_i = mole fraction corresponding to the ith component, V_t = atomic packing fraction, which was studied by Swarnakar et al. [38]

$$V_t = \rho \frac{\sum_i V_i X_i}{\sum_i M_i X_i} \tag{10.4}$$

where

M_i molar weight (ith component).

The volume of the ions (ith component) had been given by V_i. This is for a compound $M_p N_q$ given by

$$V_i = \frac{4}{3}\pi\left(pR_M^3 + qR_N^3\right)N_A \quad (10.5)$$

where

M_P and N_q metal oxides,
R radii of corresponding M and N ions,
N_A Avogadro's number.

Crack initiation and propagation was influenced by another factor that is index of brittleness. The brittleness of the prepared glasses sample had been calculated by the relation including the determined Vickers hardness H_υ, and the calculated fracture toughness K_{IC} as introduced by Lawn and Marshall [39] had determined by,

$$B = \frac{H_\upsilon}{K_{IC}} \quad (10.6)$$

Poisson's ratio (σ) and Young's modulus (E) of as-prepared glassy sample had been determined by using the relation [40] which was connected by the Poisson's ratio to the calculated packing density (V_t) of the prepared glass samples which was determined by

$$\sigma = 0.5 - \frac{1}{7.5V_t} \quad (10.7)$$

Shear modulus (S) and bulk modulus (K) were evaluated by the equation,

$$S = \frac{E}{2(1+\sigma)} \quad (10.8)$$

$$K = 1.2V_t E \quad (10.9)$$

10.4.2 Density (P) and Molar Volume (V_m)

The changes in the glass structure can be determined with the help of density. It was found in the experiment that density was increased to a smaller amount with respect to the increase in the content of Fe_2O_3 [22]. It can be caused due to the Fe_2O_3

(159.69 g/mol) addition instead of relatively lesser molecular mass of zinc oxide (81.408 g/mol). The molar volume (V_m) was calculated by following formula:

$$V_m = M_T/\rho \tag{10.10}$$

It was seen that there was a slight increase in V_m with increasing in the Fe^{3+} content. This in turn increases ion's number available per unit volume resulting in increment of V_m [22]. Oxygen packing density (OPD) is defined as the number of oxygen atoms per formula unit had been explored by the following relation:

$$\text{OPD} = (\rho \times \text{total number of oxygen atoms})/M_T \tag{10.11}$$

There had a slight increase in the OPD with increasing in Fe_2O_3 content. This clearly showed the presence of compact packing of the oxide network [22].

10.4.3 Microstructural Analysis

Li et al. [26] explored and marked the structures of the glass–ceramic specimens by the XRD method directly on the wider lateral surfaces. The diffraction peaks were examined by scanning in a 2θ-range from 10° to 80° in steps of 0.033°. The microstructures had figured by the scanning electron microscopic (SEM) observations on the same surfaces. The observed surfaces had etched with 5% hydrofluoric acid solution for 4 min and sputtered with Pt. Zhang et al. [41] examined the impact of heat treatment on microstructure of lithium disilicate glass–ceramics. The microstructural features of glass–ceramics heated at the temperature of 650 °C for different gripping times are shown in Fig. 10.1 which had been explored in their investigation. When the glass was etched by HF, lithium metasilicate (Li_2SiO_3) is more easily dissolved in 5 vol. % HF solution. All samples of lithium metasilicate crystals were shaped like dendritic. When the holding time had increased, then the dimension of crystals would be changed. At the second stage, Fig. 10.1 represents the microstructure of samples under same heat treatment at 830 °C for 3 h. Same microstructure had been observed in other samples with a slight variation of the mean size of lithium disilicate ($Li_2Si_2O_5$) crystals. Some lithium disilicate crystals in G72 had been relatively coarser and non-uniform shape.

10.4.4 Elastic Properties

Wallace et al. [42] in the year 1972 explored that the elastic properties made out of a material that pass-through stress and recovered to its original shape after releasing stress. These properties play avital part in providing important information about the bonding behavior between an isotropic nature of the bonding and structural stability

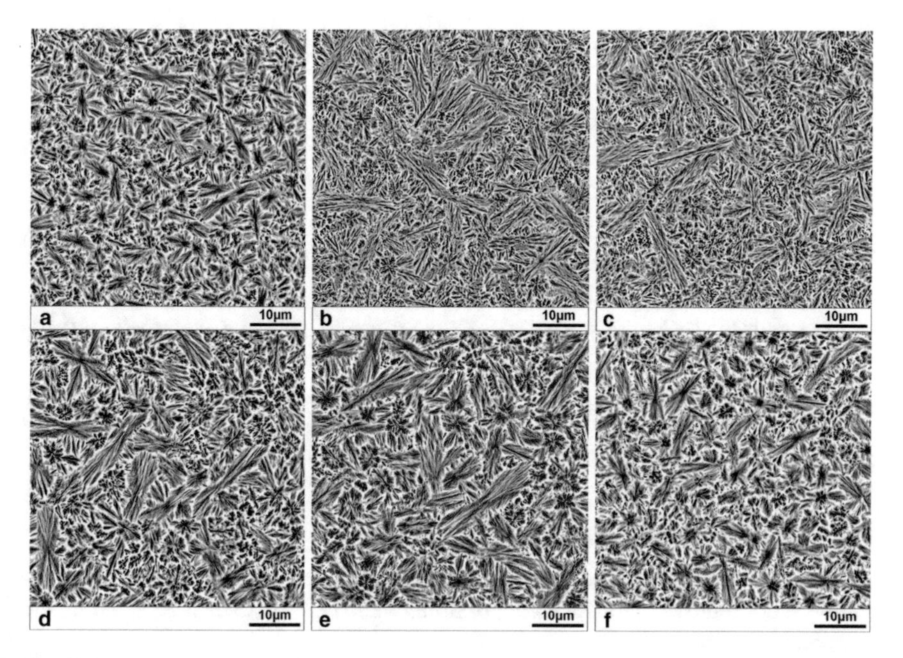

Fig. 10.1 SEM micrographs of etched surfaces of samples after the first stage treatment: **a** G3, **b** G6, **c** G12, **d** G24, **e** G48 and **f** G72. Republished with permission from Zhang et al. [41]

Fig. 10.2 A SEM image showing the representative particle sizes of the fine glass–ceramic powders for microresidual stress analysis by the XRD method. Republished with permission from Li et al. [26]

and adjacent atomic planes. There were about 21 independent elastic constants C_{ij}, but the viability of orthorhombic crystal reduced to only nine independent elastic constants mentioned (C_{11}, C_{22}, C_{33}, C_{44}, C_{55}, C_{66}, C_{12}, C_{13} and C_{23}). The elastic constants were proportional to the second-order coefficient in a polynomial fit of the total energy as a function of the distortion parameter δ. Watt et al. [43] investigated that the elastic constants had been recognized, and the other mechanical parameters would be calculated, such as Young's modulus E (GPa), shear modulus G (GPa) and Poisson's ratio ν, using the following relations:

$$G = \frac{1}{2}(G_V + G_R) \tag{10.12}$$

$$G_R = \frac{15}{4(S_{11} + S_{22} + S_{33}) - 4(S_{12} + S_{13} + S_{23}) + 3(S_{44} + S_{55} + S_{66})} \tag{10.13}$$

$$G_V = \frac{1}{15}(C_{11} + C_{22} + C_{33} - C_{12} - C_{13} - C_{23}) + \frac{1}{5}(C_{44} + C_{55} + C_{66}) \tag{10.14}$$

$$E = \frac{9BG}{3B + G} \tag{10.15}$$

$$\upsilon = \frac{3B - 2G}{2(3B + G)} \tag{10.16}$$

In the above-mentioned relations, the subscript V represents the Voigt approximation and R represents the Reuss approximation. In Eq. (10.13), the S_{ij} is the elastic compliance constants.

10.4.5 *Effect of Residual Stress*

Mastelaro et al. [44], in their investigation, they had used XRD measurements technique to elaborate the residual stresses of glass–ceramics. They have mentioned that residual stresses both tensile and compressive and also microstructure of glass–ceramics had relied on mechanical properties of glass–ceramics. Above stresses were created for the duration of the cooling from the glass transition temperature (T_g) to room temperature and were the impact of the elastic and thermal mismatch between the glass crystalline phases and glass matrix. These investigators also calculated the coefficient of linear thermal expansion for glass–ceramics which was $\alpha_g = 12.8 \times 10^{-6}/°C$ and for fully glass crystallized specimens was $\alpha_c = 10.8 \times 10^{-6}/°C$, indicating that the impacted isotropic glass crystals should be within compressive stress. XRD measurement clearly pointed out that α_c was anisotropic and relies on the glass–ceramic crystal's crystallographic orientation. For $\alpha_c > \alpha_g$, it could be told that the embedded crystals in the glass matrix remain under tension, and for $\alpha_c < \alpha_g$, it remains under compression. As reported by these researchers, the anisotropy of the coefficients of thermal expansion as well as the elastic constants of silicate crystals should be appreciating to adequately compute the residual stresses of glass–ceramics. The compressive residual stresses or tensile residual stresses in lithium disilicate glass–ceramics could be occurred, and this totally depends upon the direction of crystallographic. Soares et al. [45] utilized the indentation method to compute the residual stress of lithium disilicate glass–ceramics. The researchers had shown the experimental glass at temperature of 600 °C for various annealing process durations and observed that the residual stress level in the glass–ceramics near the crystal was high or larger, and with distance from the crystal, it gets decreased (Fig. 10.2). Residual stress could be treated as a purpose of distance from the crystal surface. When experimental specimens were heat treated for 90–120 min, then it was seen that the residual stresses declined from ~20 × 10^6 Pa to ~0 with the raising distance from the surface of crystal. The residual stress of ~ 50 × 10^6 Pa near the surface of crystal was determined when the specimen is heat treated for 240 min. The residual stresses of glass–ceramic crystals were focused in a zone, which is less than 100 μm from the surfaces of crystal. The residual stresses were minimalistic or small at larger distances of crystal surface. Residual stresses of glass–ceramic depend on the distance from the surface of crystals as well as on size of crystal which parameter was not mentioned that in this experiment. Li et al. [26], in their experiment, they have shown that the prepared glass–ceramic samples had both macro- and microresidual stresses. The macroresidual stresses impacted from the temperature gradients over the cooling that showed allocations across the samples' sections. The macroresidual stress had been evaluated by the XRD technique in steps of about 0.008°. The microresidual stresses of prepared samples precipitated by TEC mismatch had also taken into view by the XRD technique in steps of about 0.008°. By the SEM observations, it can be said that the particle sizes of the attained glass–ceramic powders had been calculated to be smaller than 1.5 μm. In the glass–ceramics powdered samples, the interactivity between the glass matrix and crystalline phase had been decreased, and the side-by-side microresidual stresses had assumed to be negligible. [26].

10.4.6 Effect of Additives

Several authors studied the effect of zirconium oxide on crystallization in lithium disilicate-based glasses. Apel et al. and their group [46] explored the changes that take place on the physical properties of lithium glass–ceramics caused by the consequence of zirconium oxide. The content of zirconium oxide varied from 0.00 to 2.010 wt.%, 2.910 and 4.050 wt.%. Viscosity of the lithium glass–ceramics increased with using maximum zirconium oxide content. For this reason, crystal growth of lithium glass has been changed and microstructure of glass–ceramics has been affected. The mobility of ions has been prevented using high viscosity of glass–ceramic and thus decreased the reaction rate of solid state on crystal-phase precipitation. There was no satisfactory relationship between content of zirconia and mechanical properties of the experimental samples. Huang et al. [47] explored the change that takes place on the properties of possible of zirconia-toughened lithium disilicate glass–ceramics when the zirconia oxide was imposed on it. Here, a hot-pressing method is utilized where the temperature is kept at 800 °C and operates within a pressure of about 30X10^6 Pa. The time taken for this is nearly 1 h, and the operation is done in vacuum condition. The main phases that were seen were as follows:

- Lithium disilicate
- Lithium metasilicate
- Tetragonal zirconium oxide (t-ZrO_2).

Investigations have been done where the characteristics of lithium disilicate glass–ceramics under the possible effect of the content of zirconium oxide are observed by a handful number of investigators. The formation of glass–ceramics as well as reaction mechanism of glass–ceramics have been influenced by the micro- and macrocrystals of zirconium oxide, $ZrSiO_4$ and precipitations of zirconium oxide-rich crystals of $Li_2ZrSiO_6O_{15}$. Hugo R. et al. [48], they have shown that in their experiment that chemical durability, densification and mechanical strength were improved by the addition of K_2O and Al_2O_3. The mechanical strength could attain up to 201.00 MPa. Nucleating agents (TiO_2 and P_2O_5) were used for formation of fine-grained interlocking microstructure in glass–ceramics, and mechanical strength was improved. This experimental investigation was done by Khater et al. [49]. The effect of the magnesium oxide additive on the properties of lithium disilicate glass–ceramics had been explored by Monmaturapoj et al. [50]. The coefficient of thermal expansion and viscosity of the molten (melting) glass are stimulated by the magnesium oxide. Phase formation and microstructure of lithium disilicate glass–ceramics were not influenced by magnesium oxide. The effects of Al_2O_3 and K_2O on the properties of lithium disilicate glass–ceramics had been explored by Fernandes et al. [51].

10.4.7 Effect of Heat Treatment

The experiment on heat treatment effect on the characteristic of lithium disilicate glass–ceramics had been explored by Zhang et al. [41]. To prepare glass–ceramics, three heat stages were used. The lithium metasilicate phase dominated in all specimens, but in case of G72, which contained small amounts of silica and lithium disilicate phase after the first stage, it was an exceptional. The crystallinity of the specimens was not affected by the heating time strongly. When heat treatment was processed for 72 h, then crystallinity was increased. The shape of the lithium metasilicate crystals was a dendritic one, and the heating time affected the size. The lithium metasilicate crystals which were about 12.39 μm were observed to be the longest in G48 and 7.56 μm in G72 to be the shortest. Metasilicate phase of lithium was absent in all specimen after the second stage of heating. The crystals of lithium disilicate resembled alike a rod shape in a closed packed and more than one directional interlocking microstructure. The mean length was found to decrease slightly from 0.4330 μm of sample G3 to 0.3220 μm of sample G48 and a small increase to 0.3960 μm of sample G72. Few crystals of lithium disilicate were coarser and eccentric shaped in G72. The lithium disilicate grain size depended on the lithium metasilicate crystals grain size. As the lithium metasilicate crystals are produced in a larger manner, lithium disilicate crystals will be finer because these lithium metasilicate crystals are provided the platform of more nucleation sites for lithium disilicate crystals.

After heat treatment in the second stage, the flexural strength was decreased continually.

The maximum mechanical strength (392 $\pm$ 27 MPa) was founded in sample of G3, and the minimum mechanical strength (242 $\pm$ 31 MPa) was founded in sample of G72. Then, the heat treatment at third stage. At 550 °C for 3 h, the bending strength was immensely upgraded that was varied from 562 $\pm$ 32 MPa to 611 $\pm$ 27 MPa. Internal stresses as well as micro-cracks were produced later the crystallization processed. For this reason, coefficient of thermal expansion as well as dissimilarity density was shown between the crystalline phase and the glass–ceramic matrix. The bending strength was reduced because of these thermal stresses and micro-cracks. In the third heat treatment, the bending strength was improved by these micro-cracks and internal stresses. Borom et al. [52] examined the results of different heating directions on the microstructure of Li_2O–Al_2 O_3–SiO_2–B_2O_3–K_2O–P_2O_5 arrangement. Then, crystal size abided adequately constant with increased in growth time. Zheng et al. [53] in their experiment used one-stage and two-stage treatments on Li_2O–SiO_2–ZnO–K_2O–P_2O_5 glassy system. Two-stage treatment [54] had been preferable for the growth of stable lithium disilicate crystals. SiO_2–Li_2O–Al_2O_3–K_2O–P_2O_5–B_2O_3 glassy system had heat treated for a sufficient time period. Burgner et al. [55] explored that the 34.5Li_2O–65.6SiO_2 (mol%) glass had been heated for over 100 h.

10.4.8 Effect of Crystal Size

The impact of crystal size on the mechanical properties of lithium disilicate glass–ceramics had been examined by Li et al. [26]. Two heating steps had been utilized for the arrangement of lithium disilicate glass–ceramics. At a constant temperature of 610 °C, the first crystallization was executed and to facilitate the crystallization of the lithium metasilicate phase. The second crystallization steps were represented at various temperatures such as 755 °C (G1), 799 °C (G2), 843 °C (G3) and 900 °C (G4) to generate the lithium disilicate phase. Figure 10.3 shows SEM images of the glass–ceramic samples: (a) G1, (b) G2, (c) G3 and (d) G4, respectively. The lithium disilicate phase exhibited a rod-like morphology which forms interlocking microstructures. The sizes of the lithium disilicate crystals were contrasting. As the annealing temperature raised, the average diameter of the crystals gets raised and the average length of the crystals raised more accurately. Therefore, the average aspect ratio of the crystals noticeably gets increased from about approx. 4 to approx. 10. Thus, it can be concluded that with the higher annealing temperature, inhibit the lithium disilicate crystals with larger aspect ratios in the glass–ceramic. Figure 10.4 represents the reliability of the flexural strength on the annealing temperature where the error bars display the standard deviations of the data. The sample G1 with smaller-sized lithium

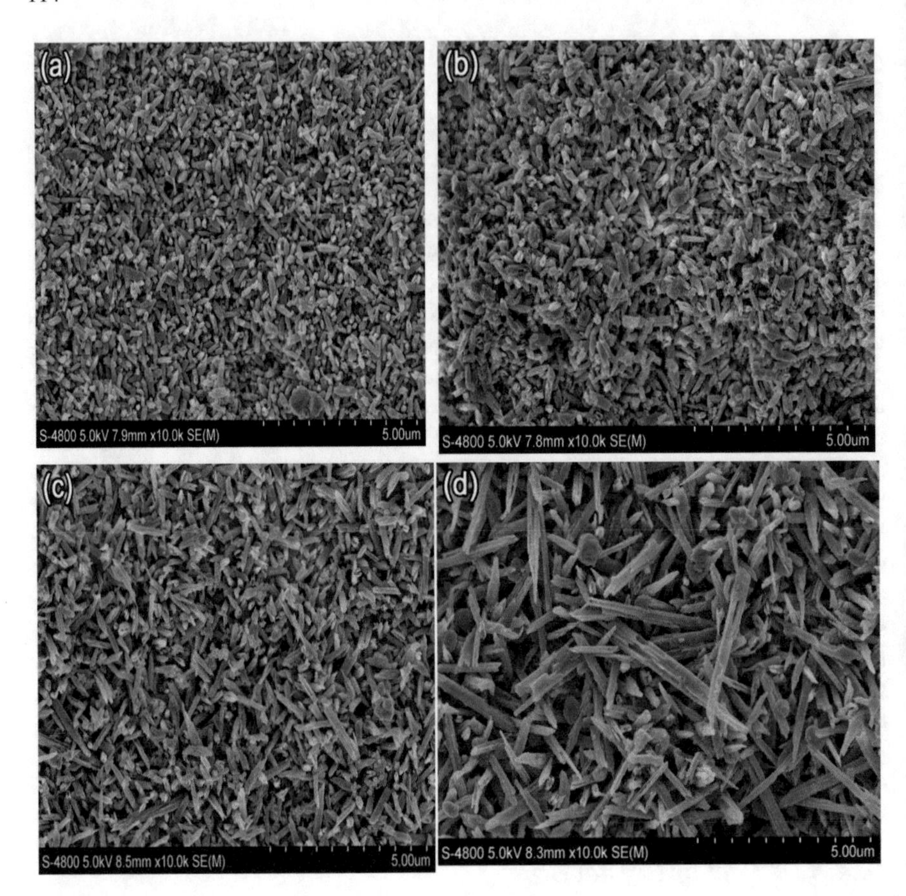

Fig. 10.3 SEM images of the glass–ceramic samples: **a** G1, **b** G2, **c** G3 and **d** G4. Republished with permission from Li et al. [26]

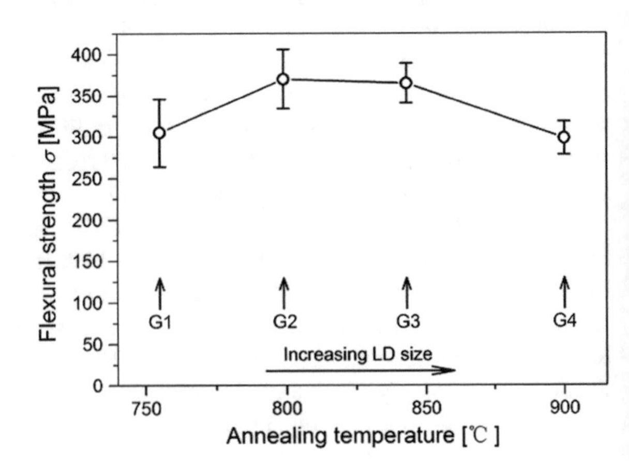

Fig. 10.4 Flexural strength (σ) of the glass–ceramic on the annealing temperature. Republished with permission from Li et al. [26]

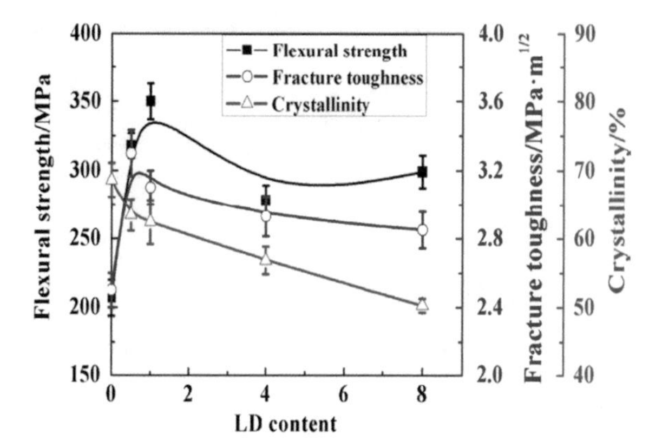

Fig. 10.5 Three-point flexural strength and fracture toughness of lithium disilicate glass–ceramics with different lithium disilicate contents. Republished with permission from Zhao et al. [56]

disilicate crystals represented lower flexural strength, which nevertheless, sample G4 with larger-sized lithium disilicate crystals also manifested lower flexural strength. There was a strength hump in the present annealing temperature range which corresponded to the precipitation of the lithium disilicate crystals with medium sizes. The flexural strength of the glass–ceramic sample did not monotonously increased with increased in the crystal size.

Zhao et al. [55] in their experiment showed that the highest crystallinity of 68.58% had been attained for the specimens of M1S2, but the minimum flexural strength and fracture toughness had seen (Fig. 10.5). High porosity nearly of about 6.02% is obtained from the inadequate glass phase. The fracture toughness as well as flexural strength of the $Li_2Si_2O_5$ glass–ceramics with the LD glass incorporation had been strongly enhanced (Fig. 10.5).

10.4.9 Translucency Effect

Zhang et al. [41] explored the effect of translucency of lithium disilicate glass–ceramics. In their experiment, they had investigated that in Fig. 10.6, the real in-line transmission (RIT) of all the glass–ceramics was shown after the third treatment stage. They had also shown that the real in-line transmission of all glass–ceramics at a wave length of 300–1100 nm showed a same changing tendency, but wavelength had increased drastically. There was a great difference at noticeable wavelengths within all the samples. The samples of G48 exhibited the maximum transmission, and its real in-line transmission at the wavelength of 650.00 nm was 29%, but in case of G3 samples, it only attained 10%.

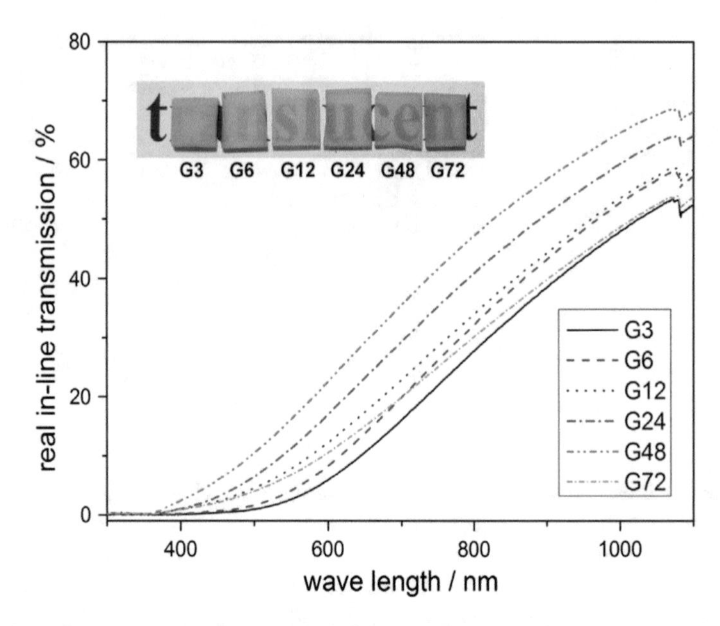

Fig. 10.6 Real in-line transmission spectra of all samples. Republished with permission from Zhang et al. [41]

10.5 Conclusion

According to this study, we have concluded that the different properties of various lithium-doped glassy materials depend on different parameters. Different types of mechanical test have been done by the following relation, like flexural strength, fracture toughness, Young's modulus, atomic packing fraction, brittleness, Poisson's ratio. With the help of microstructure, we could easily find the flexural toughness, flexural strength, elastic modulus and optical properties. Various factors such as chemical composition, additives, nucleating agent and heat treatment processes control the microstructure of lithium glassy materials. The residual stresses in lithium disilicate glass–ceramics can be compressive or tensile that depends on the crystallographic direction. In case of three stages of heat treatment process, the sample is heat treated for 48 h at the first stage and finally allowed a perfect bending strength of 581 $\pm$ 25 MPa and a highest real in-line transmission of 29% at 650 nm. Few amounts of $Li_2Si_2O_5$ and SiO_2 crystals may precipitate at the first stage, which would slightly reduce the bending strength and the translucency of finally formed glass–ceramics by the excessive extension of holding time (72 h). The chemical durability, densification and mechanical strength, which can reach up to 201 MPa, could improve on the addition of Al_2O_3 and K_2O. TiO_2 and P_2O_5 were used as efficient nucleating agents to contribute to a fine-grained interlocking microstructure in GCs, which could lead to a high mechanical strength. The crystal size accomplishes two effects on the flexural strength of the glass–ceramic such as interlocking effect and microresidual stress effect. The nano-indentation hardness depends on the microresidual stress effect. The effect of different heating routes on the microstructure of Li_2O–Al_2O_3–SiO_2–B_2O_3–K_2O–P_2O_5 system got that LD content, and crystal size remained fairly constant with increased in growth time. One-stage and two-stage heat treatments were used on Li_2O–SiO_2–ZnO–K_2O–P_2O_5 glass. Two-stage treatment was preferable for the growth of stable lithium disilicate crystals.

References

1. O. Peitl, E.D. Zanotto, F.C. Serbena, L.L. Hench, Acta Biomater **8**(1), 321 (2012)
2. F.C. Serbena, V.O. Soares, O. Peitl, H. Pinto, R. Muccillo, E.D. Zanotto, J. of the American Ceramic Society **94**(4), 1206 (2011)
3. M. Naruporn, L. Pornchanok, T. Witoon, Adv. Mater. Sci. Eng. (2013)
4. D.R. Stookey, Indus. Eng. Chem. Res. **51**, 805 (1959)
5. G.H. Beall, Am. Ceram. Soc. **251** (1971)
6. S.W. Freiman, L.L. Hench, J. Am. Ceram. Soc. **55**, 86 (1972)
7. A.C. Hannon, B. Vessal, J.M. Parker, J. Non-Cryst. Solids **150**, 97 (1992)
8. P.C. Soares Jr., E.D. Zanotto, V.M. Fokin, H. Jain, J. Non-Cryst. Solids **331**, 217 (2003)
9. H. Saifang, Li Ying, W. Shanghai, H. Zhaohui, G. Wei, C. Peng, J. Euro. Ceram. Soc. **137**, 1083 (2017)
10. E.B. Zine, R. Habib, B. Merzoug, R. Djamel, J. Mech. Behav. Biomed. Mater. **32**, 345 (2014)
11. D. Binghui, L. Jian, T. Harris Jason, M. Smith Charlene, E. McKenzie Matthew, J. Am. Ceram. Soc. **1** (2019)
12. A. Fukabori, T. Yanagida, V. Chani, F. Moretti, J. Pejchal, Y. Yokota, N. Kawaguchi, K. Kamadaa, K. Watanabe, T. Murata, Y. Arikawa, K. Yamanoi, T. Shimizu, N. Sarukura, M. Nakai, T. Norimatsu, H. Azechi, S. Fujinoi, H. Yoshida, A. Yoshikawa, J. Non-Cryst. Solids **357**, 910 (2011)
13. L. Hallmanna, P. Ulmer, K. Mathias, J. Mech. Behav. Biomed. Mater. (2018)
14. M. Chatterjee, M.K. Naskar, Ceram. Int. **32**, 623 (2006)
15. B. Karmakar, P. Kundu, S. Jana, R.N. Dwivedi, J. Am. Ceram. Soc. **85**(10), 2572 (2002)
16. A. Arvind, R. Kumar, M.N. Deo, V.K. Shrikhande, G.P. Kothiyal, Ceram. Int. **35**, 1661 (2009)
17. W. Hölland, V. Rheinberger, E. Apel, C. Hoen, J. Euro. Ceram. Soc. **27**, 1521 (2007)
18. J.B. Quinn, V. Sundar, I.K. Lloyd, Dent. Mater. **19**, 603 (2003)
19. W. Hölland, G. Beall, Glass-ceramic technology. Am. Ceram. Soc. (2002)
20. B. Li, Z. Qing, Y. Li, H. Li, S. Zhang, J. Mater. Sci. Mater Electron **27**, 2455 (2016)
21. B. Indrajit Sharmaa, M. Goswamia, P. Senguptab, V.K. Shrikhandea, G.B. Kaleb, G.P. Kothiyala, Mater. Lett. **58**, 2423 (2004)
22. H.D. Subhashini, N.K. Shashikala, U. Shankar, Ceram. Int. (2019)
23. M.H.A. Mhareb, S. Hashim, S.K. Ghoshal, Y.S.M. Alajerami, M.A. Saleh, R.S. Dawaud, N.A.B. Razak, S.A.B. Azizan, Opt. Mater. **37**, 391 (2014)
24. S.M. Salman, H. Darwish, E.A. Mahdy, Mater. Chem. Phys. **112**, 945 (2008)
25. L. Hallmann, P. Ulmer, M. Kern, J. Mech. Behav. Biomed. Mater. **82**, 355 (2018)
26. D. Li, J.W. Guo, X.S Wang, S.F. Zhang, L. He, Mater. Sci. Eng. (2016)
27. X. Huang, X. Zheng, G. Zhao, B. Zhong, X. Zhang, G. Wen, Mater. Chem. Phys. **143**(2), 845 (2014)
28. G. Wen, X. Zheng, L. Song, Acta Mater **55**(10), 3583 (2007)
29. I. Denry, J. Holloway, Materials **3**(1), 351 (2010)
30. S. Buchner, C.M. Lepienski, P.C. Jr Soares, N.M. Balzaretti. Mat. Sci. Eng. **528**(10) 3921 (2011)
31. P. Goharian, A. Nemati, M. Shabanian, A. Afshar, J. Non-Cryst. Solids **356**, 208 (2010)

32. W. Holland, M. Schweiger, M. Frank, V. Rheinberger, J. Biomed. Mater. Res. **53**, 297 (2000)
33. Sang-Chun, J. Korean Acad. Prosthodont. **40**, 572 (2002)
34. I.L. Denry, J.A. Holloway, Dental Mater **20**, 213 (2004)
35. International Organization for Standardization, ISO 6872 (2008) *Dentistry–Ceramic Materials*, 3rd edn
36. G.R. Anstis, Transformation **46**, 533 (1981)
37. A. Makishima, J.D. Mackenzie, J. Non-Cryst. Solids **12**, 35 (1973)
38. A.K. Swarnakar, A. Stamboulis, D. Holland, O. Van der Biest, J. Am. Ceram. Soc. **96**, 1271 (2013)
39. B.R. Lawn, D.B. Marshall, J. Am. Ceram. Soc. **62**, 347 (1979)
40. A. Makishima, J.D. Mackenzie, J. Non-Cryst. Solids **22**, 305 (1976)
41. P. Zhang, X. Li, J. Yang, S. Xu, J. Non-Cryst. Solids **402**, 101 (2014)
42. D.C. Wallace, *Thermodynamics of Crystals* (Wiley, New York, 1972)
43. J.P. Watt, J. Appl. Phys. **50**, 6290 (1979)
44. V.R. Mastelaro, E.D. Zanotto, J. Non-Cryst. Solids **194**, 297 (1996)
45. P.C. Soares, C.M. Lepienski, J. Non-Cryst. Solids **348**, 139 (2004)
46. E. Apel, C. Hoen, V. Rheiberger, W. Höland, J. Euro. Ceram. Soc. **27**, 1571 (2006)
47. X. Huang, X. Zheng, G. Zhao, B. Zhong, X. Zhang, G. Wen, J. Mater. Chem. Phys. **143**, 845 (2014)
48. H.R. Fernandes, D.U. Tulyaganov, I.K. Goel, J.M.F. Ferreira, J. Am. Ceram. Soc. **91**, 3698 (2008)
49. G.A. Khater, M.H. Idris, Ceram. Int. **33**, 233 (2007)
50. N. Monmaturapoj, P. Lawita, W. Thepsuwan. Adv. Mater. Sci. Eng. **1** (2013)
51. H.R. Fernandes, D.U. Tulyaganov, A. Goel, M.J. Ribeiro, M.J. Pascual, J.M.P. Ferreira, J. Euro. Ceram. Soc. **30**, 2017 (2010)
52. M.P. Borom, A.M. Turkalo, R.H. Doremus, J. Am. Ceram. Soc. **58**, 385 (1975)
53. X. Zheng, G. Wen, L. Song, X.X. Huang, Acta Mater. **56**, 549 (2008)
54. H.L. Mc Collister, S.T. Reed, Glass ceramic seals to inconel, U.S. Patent 4414282 A (1983)
55. L.L. Burgner, P. Lucas, M.C. Weinberg, P.C. Soares Jr., E.D. Zanotto, J. Non-Cryst. Solids **274**, 188 (2000)
56. T. Zhao, Y. Qin, P. Zhang, B. Wang, J.F. Yang, Ceram. Int. (2014)

Chapter 11
Thermal Properties of Some Li-Doped Glassy Systems

Shayeri Das and Sanjib Bhattacharya

Abstract In the present review, the recent progress in unfolding the complexities of thermal properties of all types of Li-doped glassy composites has been comprehensively examined. The effectiveness of thermal properties for the intrinsic characterization of glassy systems and their composites was clearly demonstrated in this research. Furthermore, the utility of the thermogravimetric analysis employed for thermal characterization that has been reported by various researchers was exhaustively analyzed in this paper. This research primarily focused on the analyses of several good articles concerned with Li-doped glassy systems to assess its effect on the thermal properties of its corresponding composites. Such systematic analysis of previous literatures divulged a direction to the researchers about the solution of upgraded interfacial properties. This current research has suggested that the presence of the Li causing changes in thermal properties. The potential applications, current challenges and future perspectives pertaining to these systems have been ostentatiously discussed in the current study with regard to the promising developments of the glassy systems.

11.1 General Consideration

A glassy state is a classic instance of non-equilibrium metastable one [1]. Transition of glass from supercoiled liquid to a glassy state is viewed as a variation of relaxation phenomena [1]. During the relaxation processes, a physical quantity is contingent on time or frequency [1]. For the purpose of experimental studies of the relaxation, complex dielectric constants have been thoroughly studied. However, it has been found that the dielectric relaxation relates only to polar atomic motions. Elastic constants have been also studied in various kinds of glass-forming materials, while

S. Das
Department of Electrical Engineering, Ideal Institute of Engineering, Kalyani Shilpanchal, Kalyani, Nadia, West Bengal 741235, India

S. Bhattacharya (✉)
UGC-HRDC (Physics), University of North Bengal, Darjeeling, West Bengal 734013, India
e-mail: sanjib_ssp@yahoo.co.in; ddirhrdc@nbu.ac.in

S. Bhattacharya and K. Bhattacharya (eds.), *Lithium Ion Glassy Electrolytes*,
https://doi.org/10.1007/978-981-19-3269-4_11

the elastic relaxation relates only to density fluctuations. Concerning to lithium-doped glasses, in particular, it is quite challenging to execute the experiment of dielectric relaxation because of the ionic conductivity of lithium ion of the system.

Variable glasses have attracted the attention of various researchers in the recent times due to their wide application range [2–6]. Lithium-doped glasses containing rare earth ions are used as proficient lasing materials [7]. On introduction of alkali halides into the glass structure, they demonstrate their presence by altering the glass structure [8–11], which has been verified with the help of a number of experimental techniques. Differential scanning calorimetry (DSC) is one of the important techniques to study the glass-forming ability of a melt which in turn be governed by the cooling rate. The difference in the glass transition temperature (T_g) and crystallization temperature (T_c) is associated with the stability of the glass structure [12]. T_g is dependent upon the coordination number of the network forming ions which are a part of network forming atoms and the number of non-bridging oxygen atoms (NBOs) [13]. Consequently, a variation in T_g is always accredited to the variations befalling in the glass structure. A diminution in T_g specifies the decrease in oxygen packing density and that the structure becomes slackly packed [14]. Intended for the preparation of glass–ceramic materials, the knowledge of thermal behavior of parent glasses, and the mechanism of crystallization is very valuable. Some basic thermal parameters as glass transition temperature, dilatation softening temperature, thermal expansion coefficient, crystallization temperature, etc., are useful to know nucleation rate under controlled heat treatment.

Similarly, the impact of Li_2O and ZnO oxides in multiple component-based tellurite glasses has been investigated in different systems like: $xLi_2O–(100-x)$ $[0.25ZnO–0.15B_2O_3–0.60TeO_2]$ $x = 0, 5, 10, 15$ and 20 [15], $(80-x)$ $TeO_2–xLi_2O–20ZnO$ with $x = 0$, 5 and 10 (in mol %) [16] and $60TeO_2–15V_2O_5–(25–x)$ $ZnO–xLi_2O$ [17]. For these opuses, both oxides undertake the responsibility of network modifier. This feature escalates the glass-forming range, but is a shortcoming in optical properties: The optical absorption edge (band gap) formed between HOMO–LUMO layers, and refractive index can decrease with the addition of these oxides for the aforesaid systems. Correspondingly, the change in refractive index has direct relation with the dielectric constant of the system. The dielectric constant is a key parameter in designing of the electronic devices. In addition, it plays an important role to understand the behavior of charge carriers, dopants, defects and impurities in insulators and semiconductors [18].

An update on the research pertaining to glassy systems, dealing extensively with the thermal properties of Li^+-doped glassy systems, has been discussed in this study. The study mainly focuses on the fundamental properties of glassy systems and its diverse properties including thermal conductivity, coefficient of thermal expansion, etc. Additionally, various researchers have abridged research outcomes of recent experimental studies conducted on Li-based glassy systems for the evaluation of thermal properties. The aforementioned properties of hybrid composites have also been amalgamated and conferred in brief. Research gaps have been well identified as best as possible, and directions for future work are also provided. Concluding data has been presented in this research to illustrate the fabrication processes of glassy systems for the thermal properties of the systems.

11.2 Different Thermal Properties of Some Chalcogenide Glassy Systems (Literature Survey)

DSC thermograms for the various samples are depicted in Fig. 11.1 [19]. The researchers have observed two glass transition temperatures T_{g1} and T_{g2} for individual samples [19–21]. These transitions have been reported to be moderately weak as compared to the conventional transitions reported till date [20]. Hence, the reported glasses [19] have been categorized as "soft glasses." It has also been observed a continuous increase in the value of T_{g1} with initiation and then escalation in the concentration of Bi_2O_3 except for a particular sample with $x = 15$. The decline in T_{g1} has been attributed to the glass-forming behavior of the sample. It not only acted as a modifier, but also as a glass former. According to Arora et al. [19], Bi_2O_3 after entering in the glass matrix induces new species to be molded due to change in oxygen packing density. The new species includes BO_2F units whose formation is reported earlier by the same group of researchers [21] using the technology of FT-IR spectroscopy. The process seems to make the structure relatively strongly packed, thus increasing the value of T_{g1}. Alternatively, small changes in the values of T_{g2} have been noted with increase in Bi_2O_3 concentration.

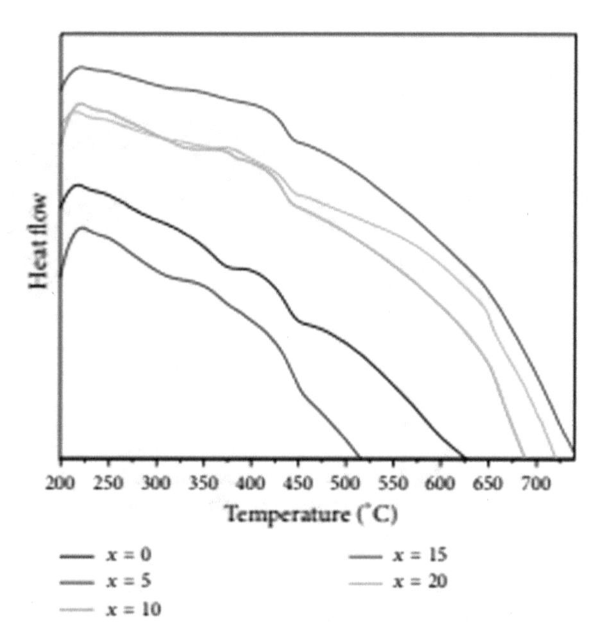

Fig. 11.1 DSC of glasses with Bi_2O_3 with Li dopants. Republished with permission from Arora et al. [19]

Karunakaranet al. [22] have reported that DSC thermograms of all lithium flruoborate glasses doped with Dy^{3+} as shown in Fig. 11.2. The glass samples were subjected to heat treatment in nitrogen environment at 10 °C/min in the temperature range of 30–1200 °C. The smooth and broader exothermic peaks and the next endothermic peaks in the DSC thermograms designate crystallization temperature (T_c) and melting temperature (T_m) [23]. The deficiency of sharp endothermic and exothermic peaks designates the homogeneous formation of the glasses [24]. From the DSC profile of the Dy^{3+}-doped lithium fluroborate glasses, the glass transition temperature (T_g), crystallization temperature (T_c) and melting temperature (T_m) [22] had been identified. In addition to these values, the glass stability factor (S) and Hruby's parameter (H_R) have also estimated and analyzed by the group of researchers [22–24]. Glasses exhibiting a high thermal stability ($T_c - T_g$) and low-temperature interval ($T_m - T_c$) have come up as the best candidates for the fiber fabrication due to the compatibly small chance of crystallization problems [25]. The pertinent parameter that summarizes both these characteristics is the Hruby parameter (H_R) defined as $(T_c - T_g)/(T_m - T_c)$ [24]. It is anticipated for a glassy host to have S as large as possible to accomplish a large viable range of temperature during glass sample fiber drawing [25]. For all such glasses, the approximate glass thermal stability is higher than 100 °C, and hence, these glasses are suitable for fiber drawing.

Sreenivasulu et al. [26] have designed stoichiometric mixture of TeO_2 (sigma Aldrich-99þ purity), ZnO (merck-99% purity), CdO (merck-99% purity) and Li_2O (merck-99.9% purity) chemicals, which have been made by melt quenching technique. DSC has been performed on the prepared glassy sample from the aforementioned mixture. DSC thermogram, shown in Fig. 11.3, demonstrates three steps in the

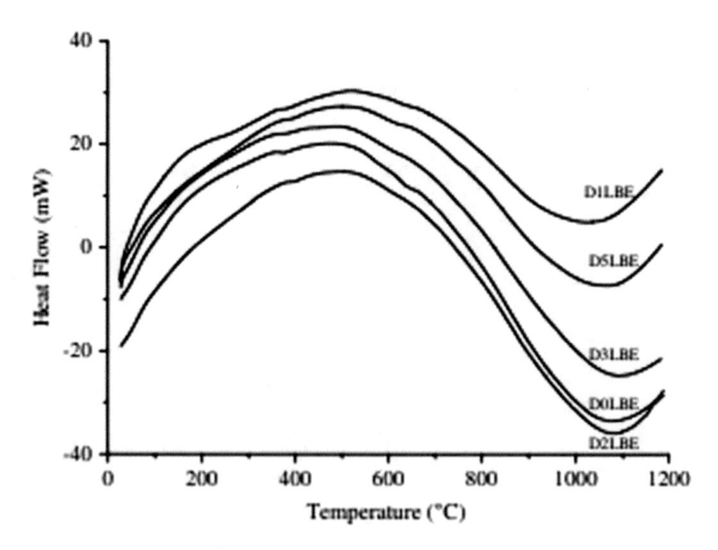

Fig. 11.2 DSC curves of the Dy^{3+}-doped lithium fluroborate glasses. Republished with permission from Karunakaran et al. [22]

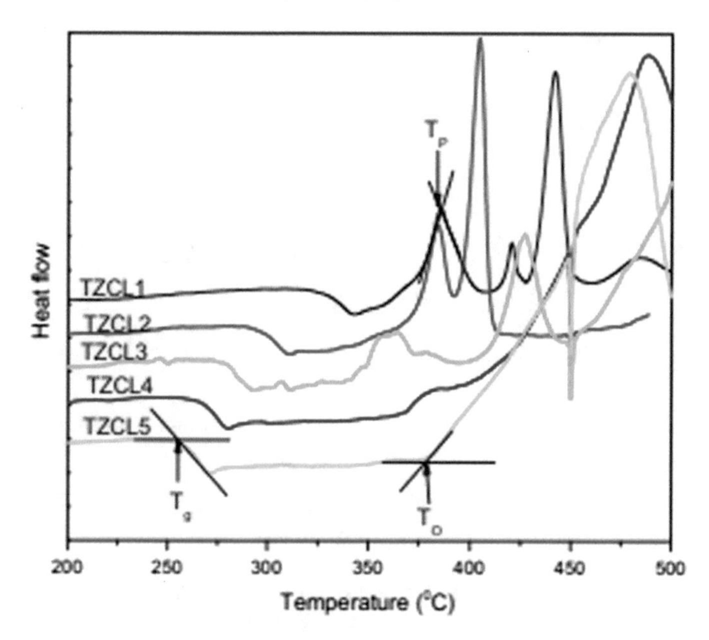

Fig. 11.3 DSC of lithium cadmium zinc tellurite glass system

graph analogous to the glass transition temperature (T_g), the onset crystalline temperature (T_o) and crystallization temperature (T_p), respectively [26]. These values are have been further analyzed by them, and it has been observed from Fig. 11.3 that T_g has decreased from 326 to 255 °C with the addition of Li_2O content from 0 to 20 mol%. The decline in T_g is attributed to a decrease in density and OPD of the glass network [26]. In accordance with literature [27, 28], it is well acknowledged that ($T_o - T_g$) is a measure of thermal stability (ΔT) of such glasses. These values have been determined and noted that thermal stability of glasses has increased with increase of Li_2O content, and it specifies that the process of crystallization is delayed in comparison with CdO content, thereby resulting in increase of the thermal stability of the present glasses.

Novatski et al. [29] have developed a glassy system, (100–*X*–*Y*) TeO_2–XLi_2O–YZnO with $X = 10, 15$ and 20, $Y = 0, 5, 10, 15, 20, 25$ and have performed DSC scans for all compositions. One such DSC scan for 30 mol% is presented in Fig. 11.4. From the observed experimental data, the calculation of the glass transition temperature (T_g) and the inception crystallization (T_x) [29] was conducted by the intersection point amid baseline and tangent line in inflection point at the first change in this reference point for (T_g) and the inflection point of endothermic peaks for (T_x). These values have been computed, and it has been observed that T_g has the similar pattern for the three groups of samples and residues constant with the substitution of TeO_2 for ZnO [30]. It is well recognized that the addition of ZnO reduces T_g values due to the creation of non-bridging oxygens (NBO), which is a factor that delineates the ZnO as a modifier network oxide [31]. Nevertheless, for the TeO_2–Li_2O–ZnO

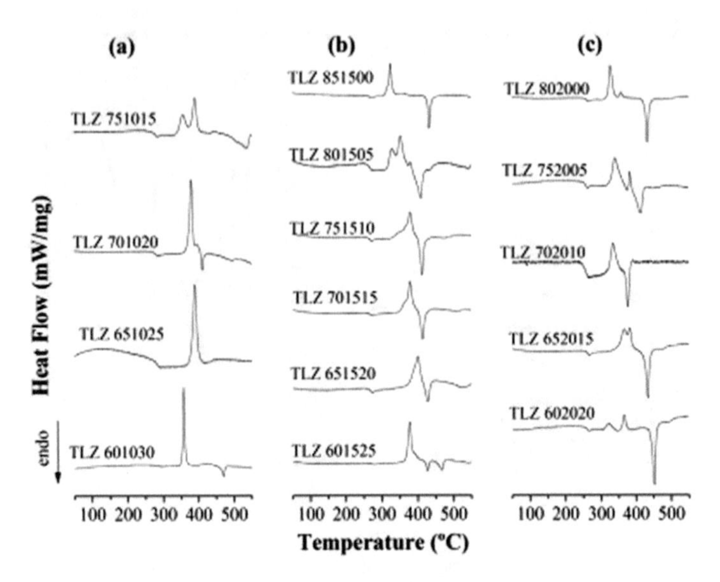

Fig. 11.4 DSC scans of samples: **a** GL10, **b** GL15 and **c** GL20. T_g values remain constant for each group. Comparing the three groups, GL20 presents the lowest T_g values (average). ΔT increases up to $TeO_2 = 65$ mol% for the three groups. Republished with permission from Novatski et al. [29]

system, this is not perceived, due to the influence of Li_2O [29]. However, the addition of Li_2O creates a substantial amount of NBOs due to the breaking of TeO_4 linkage [31], and it could be suggested that the ZnO addition inhibits the effects of Li_2O. Associating the three groups of samples, it has been cleared that the GL20 portrays the average lowest T_g values accentuating the role of Li_2O as a network modifier on the studied system [31]. Additionally, this behavior occurs to samples in which Li_2O contented is higher than ZnO, reinforcing the statement that ZnO addition escalates the network connectivity.

Further studies have been conducted by various other researchers [16, 32–35] on melting temperature, glass transition temperature, thermal conductivity and thermal expansion coefficient of other glass–nanocomposites containing Li_2O for assimilation of thermal properties of them, which is expected to become a pivotal role for proper understanding of their structure.

11.3 Conclusions

The thermal properties of Li-doped glassy systems have been thoroughly reviewed. The entire discussion of the thermal properties has revealed that Li^+ plays an important role in the overall properties of glassy systems. It has been observed repeatedly that some clusters with decreasing content also decrease density and molar volume [29]. Similarly, it has also been perceived that the thermal stability of glasses increased with the increase in Li_2O content [26]. Calorimetric T_g and the width of the glass transition region have been resoluted as a function of Li_2O. They illustrate the remarkable changes with the increase of composition. This additive effect of has led to the structural change of the glass system and prompts the increase of the fragility markedly, which correlated the width of glass transition region [1]. This review contains an analysis on the research investigations which have been exhaustively performed for assessing the thermal properties of glassy systems which shows the scope for further research.

References

1. Y. Matsudaa, C. Matsuib, Y. Ikea, M. Kodamac, S. Kojimaa, AIP Conf. Proc. **832**, 155 (2006)
2. T. Catunda, M.L. Bacsso, Y. Menaddeeq, M.A. Aegester, J. Non-Cryst, Solids **213**, 225 (1997)
3. M.R. Ozalp, G. Ozwn, A. Sennaroghu, A. Kurt, Opt. Commun. **217**, 281 (2003)
4. Y. Ohisi, M. Yamda, J. Kanamori, S. Suda, Opt. Lett. **22**, 1235 (1997)
5. M.C. Brierely, C.A. Miller, Electron. Lett. **24**, 438 (1988)
6. A. Lecoq, M. Poulain, J. Non-Cryst, Solids **41**, 209 (1980)
7. V.R. Kumar, N. Veeraiah, B.A. Rao, J. Lumin. **75**(1), 57–62 (1997)
8. S. Arora, V. Kundu, D.R. Goyal, A.S. Maan, ISRN Optics **2012**, 7 (2012). Article ID 193185
9. N. Soga, J. Non-Cryst. Solids **73**(1–3), 305–313 (1985)
10. I.Z. Hager, J. Mater. Sci. **37**(7), 1309–1313 (2002)
11. I.Z. Hager, M. El-Hofy, Physica Status Solidi A **198**(1), 7–17 (2003)
12. S. Hazra, S. Mandal, A. Ghosh, Phys. Rev. B **56**(13), 8021–8025 (1997)
13. M. Subhadra, P. Kistaiah, Physica B **406**(8), 1501–1505 (2011)
14. N.J. Kim, Y.H. La, S.H. Im, W.T. Han, B.K. Ryu, Electron. Mater. Lett. **5**(4), 209–212 (2009)
15. S. Rani, N. Ahlawat, R. Parmar, S. Dhankhar, R.S. Kundu, Indian J. Phys. 1–9 (2008)
16. E.A. Mohamed, F. Ahmad, K.A. Aly, J. Alloy. Comp. **538**, 230–236 (2012)
17. S. Rani, R.S. Kundu, N. Ahlawat, S. Rani, K.M. Sangwan, N. Ahlawat, in *AIP Conference Proceedings* (AIP Publishing, 2018), p. 70026
18. A.S. Verma, D. Sharma, Phys. Scr. **76**, 22 (2007)
19. S. Arora, V. Kundu, D.R. Goyal, A.S. Maan, Hindawi Publication **914324**, 5 (2013)
20. M.H. Shaaban, J. Mater. Sci. **47**, 5823–5832 (2012)
21. S. Arora, V. Kundu, D.R. Goyal, A.S. Maan, ISRN Spectrosc. **2012**, 5. Article ID 896492 (2012)
22. R.T. Karunakaran, K. Marimuthu, S. SurendraBabu, S. Arumugam, Physica B **404**, 3995–4000 (2009)
23. W.T. Carnall, P.R. Fields, K. Rajnath, J. Chem. Phys. **49**, 4424 (1968)
24. A. Hruby, Czech. J. Phys. B **22**, 1187 (1972)
25. Z. Pan, S.H. Morgan, J. Non-Cryst, Solids **210**, 130 (1997)
26. V. Sreenivasulu, G. Upender, Swapna, V. VamsiPriya, V. Chandra Mouli, M. Prasad, Physica B **454**, 60–66 (2014)
27. Y. Li, S.C. Ng, Z.P. Lu, Y.P. Feng, K. Lu, Philos. Mag. Lett. **78**(3), 213–220 (1998)
28. A. Bajaj, A. Khanna, J. Phys. **21**, 8. Article ID 035112 (2009)
29. A. Novatskia, A. Somera, A. Gonçalvesa, R.L.S. Piazzettaa, J.V. Gunhaa, A.V.C. Andradea, E.K. Lenzia, A.N. Medinab, N.G.C. Astrathb, R. El-Mallawanyc, Mater. Chem. Phys. **231**, 150–158 (2019)
30. H.A.A. Sidek, S. Rosmawati, B.Z. Azmi, A.H. Shaari, Adv. Condens. Matter Phys. (2013)
31. T. Sekiya, N. Mochida, A. Ohtsuka, M. Tonokawa, J. Non-Cryst, Solids **144**, 128–144 (1992)
32. S. Manning, H. Ebendorff-Heidepriem, T.M. Monro, Opt. Mater. Express **2**, 140–152 (2012)

33. R. El-Mallawany, J. Mater. Sci. Mater. Electron. **6**, 1–3 (1995)
34. J.L. Gomes, A. Gonçalves, A. Somer, J.V. Gunha, G.K. Cruz, A. Novatski, J. Therm. Anal. Calorim. **134**, 1439–1445 (2018)
35. T. Kosuge, Y. Benino, V. Dimitrov, R. Sato, T. Komatsu, J. Non-Cryst, Solids **242**, 154–164 (1998)

Chapter 12
Comparison Between Some Glassy Systems and Their Heat-Treated Counterparts

Aditi Sengupta, Chandan Kr Ghosh, and Sanjib Bhattacharya

Abstract Li_2O-doped glass-nanocomposites and crystalline counterparts have been developed. Microstructural study reveals the distribution of $Li_2Zn_2(MoO_4)_3$, $ZnMoO_4$, $Zn(MoO_2)_2$, $Li_2Mo_6O_7$ and Li_2MoO_3 nanorods in the glassy matrices. Crystalline counterparts exhibit enhancement in crystallites sizes. The ionic conductivity is found to be function of frequency and temperature. Flat conductivity at a low-frequency regime indicates the diffusional motion of Li^+, whereas the "higher frequency dispersion" may correspond to a correlated and sub-diffusive motion. As the crystalline counterpart is formed by controlled heating, $ZnSeO_3$ chain structure is expected to break by increasing dimensions of molybdate rod-like structures.

Keywords Amorphous materials · Ceramic composites · Electrical properties · Microstructure

12.1 Introduction

Physical and electrical properties of nanophased materials are of great scientific interest in present time [1]. The improved electrical properties of nanophased materials in comparison with the bulk counter parts are main reason of the present work. Comparing the electrical and mechanical properties of the conventional materials so far available, efforts would be made towards the evolution of a better material suitable for engineering practice [2]. Comparative study on electrical properties like

A. Sengupta
Department of Electronics and Communication Engineering, Siliguri Institute of Technology, Siliguri, Darjeeling, West Bengal 734009, India

C. K. Ghosh
Department of Electronics and Communication Engineering, Dr. B. C. Roy Engineering College, Durgapur 713026, West Bengal, India

S. Bhattacharya (✉)
UGC-HRDC (Physics), University of North Bengal, Darjeeling 734013, West Bengal, India
e-mail: sanjib_ssp@yahoo.co.in; ddirhrdc@nbu.ac.in

S. Bhattacharya and K. Bhattacharya (eds.), *Lithium Ion Glassy Electrolytes*,
https://doi.org/10.1007/978-981-19-3269-4_12

ionic conduction of some oxide glassy nanocomposites with their crystalline counterparts is the key interest of this chapter, which must contain the information about the most likely migration pathways [1–3]. This can be achieved by exploring the structural differences of them, which may cause to change in ion dynamics in present glassy systems and their crystalline counterparts. The difficulties to characterize the structure (specially the defects in the structure) of an amorphous material create challenges in understanding them than that of a crystalline material [3].

12.2 General Consideration

Lithium is supposed to be one of the most promising candidates of rechargeable battery electrolytes for its role in electric vehicles, mobile computers, etc., as well as for academic interest [1]. Some short falls of conventional lithium-ion batteries [1]have been already identified due to highly flammable nature of organic liquid electrolytes or polymer electrolytes. Researchers instigate some safety issues to develop "solid electrolytes" for their high thermal stability, high energy density and better electrochemical stability [2]. "Se" may consist of long polymeric chains with ring fragments as a mixture of Se_n chains and Se_8 rings [3]. So, the change in microstructure of Li_2O-doped glassy ceramics and their crystalline counterparts may play important role in conduction process.

To understand the mechanism of conductivity spectra and relaxation process in lithium ion conductor, Jonscher's power law model [4–6] should be unique in general, which describes the total conductivity as:

$$\sigma(\omega) = \sigma_o + A\omega^S \tag{12.1}$$

where A is the prefactor, S is the frequency exponent and σ_o is the low-frequency conductivity. In particular, the theoretical form of conductivity spectra is proposed by Almond-West formalism [4–6],

$$s(w) = s_{dc}[1 + (w/w_H)^n] \tag{12.2}$$

which is the combination of the DC conductivity (σ_{dc}), hopping frequency (ω_H) and a fractional power law exponent (n). To acquire sufficient information regarding AC conduction [4–6], this model can be successfully exploited to fit the experimental data.

Investigation of electrical transport [7] of Li_2O-doped glass ceramics and heat-treated counterparts is expected to reveal structural information, which is essential for the assimilation of conduction mechanism. A comparison of conduction behavior of Li_2O-doped new glass ceramics and crystalline counterparts has been studied to explore new features. The effect of Li_2O doping in the glassy ceramics and crystalline counterparts is expected to be interesting not only for technological applications but also for academic interest.

12.3 Various Cases

12.3.1 Laboratory Experiment and Measurement of As-Prepared Glassy Samples

A series of glass-nanocomposites, xLi_2O–$(1-x)$ ($0.05ZnO$–$0.475MoO_3$–$0.475SeO_2$) with $x = 0.1$ and 0.3 are prepared by melt quenching at ~700 °C for 30 min. Second set of samples (mixtures) are allowed to pass through a slow cooling during 17 h. The final product is obtained in the form of a solid sample (crystalline) without quenching. The electrical measurements of them are performed using a high-precision LCR meter (HIOKI, model no. 3532–50) in a wide temperatures and frequencies (42–5 MHz) ranges. Scanning electron microscopic (SEM) pictures and X-ray diffractograms are analyzed to get microstructures.

12.3.2 Results and Discussion

The XRD patterns of the present glassy system and crystalline counterparts are presented in Fig. 12.1a, which exhibits more crystallinity for $x = 0.3$ due to formation of sharp peaks [7]. Formation of $ZnSeO_3$, $Li_{10}Zn_4O_9$, $Li_2Zn_2(MoO_4)_3$, $Li_{10}Zn_4O_9$, $ZnMoO_4$, $Zn(MoO_2)_2$, $Li_2Mo_6O_7$, Li_2MoO_3 and $Li_{10}Zn_4O_9$ nanophases [7] is confirmed from XRD study. Average crystallite sizes estimated from Scherer relation [7] are presented in Fig. 12.1b. Crystallite sizes are found to increase for glassy system as well as crystalline counterparts. This result requires structural modifications of the present system. Figures 12.1c, d exhibit SEM micrographs for $x = 0.1$ and crystalline counterpart, which shows surface morphology with small irregular grains as well as rod-like structures. As molybdenum tends to form a rod-like structure [7], $Li_2Zn_2(MoO_4)_3$, $ZnMoO_4$, $Zn(MoO_2)_2$, $Li_2Mo_6O_7$ and Li_2MoO_3 nanorods of average length 5 μm and breadth 50 nm are formed in glassy matrices for $x = 0.1$. The dimensions of such nanorods are found to increase in crystalline counterparts as illustrated in Fig. 12.1d. As the selenite compounds are found to associate with the formation of disorder in the SeO_3 chains by suitable modifier [7], $ZnSeO_3$ nanostructures in the form of interconnected chains are expected to develop in the present system. In the crystalline counterparts, preferred heat treatments may lead to break this chain structure by increasing the length and breadth of molybdate rod-like structures as exhibited in Fig. 12.1d. Formation of more voids (defects) with trapping of Li^+ ions may be the possible outcomes of these results. In addition, $Li_2Zn_2(MoO4)_3$ crystal-melt is formed from MoO_3 and Li_2MoO_4–$ZnMoO_4$ [7], where MoO_3 acts as a solvent with higher volatility is compared to Li_2O and ZnO. Transformation of Li_2MoO_3 phase [7] from layered into disordered cubic structure is expected to favorable in conduction process.

DC electrical conductivity can be extracted from the complex impedance plots [5–7], which are shown in Fig. 12.2a for 473 K. All the semicircular arcs show different

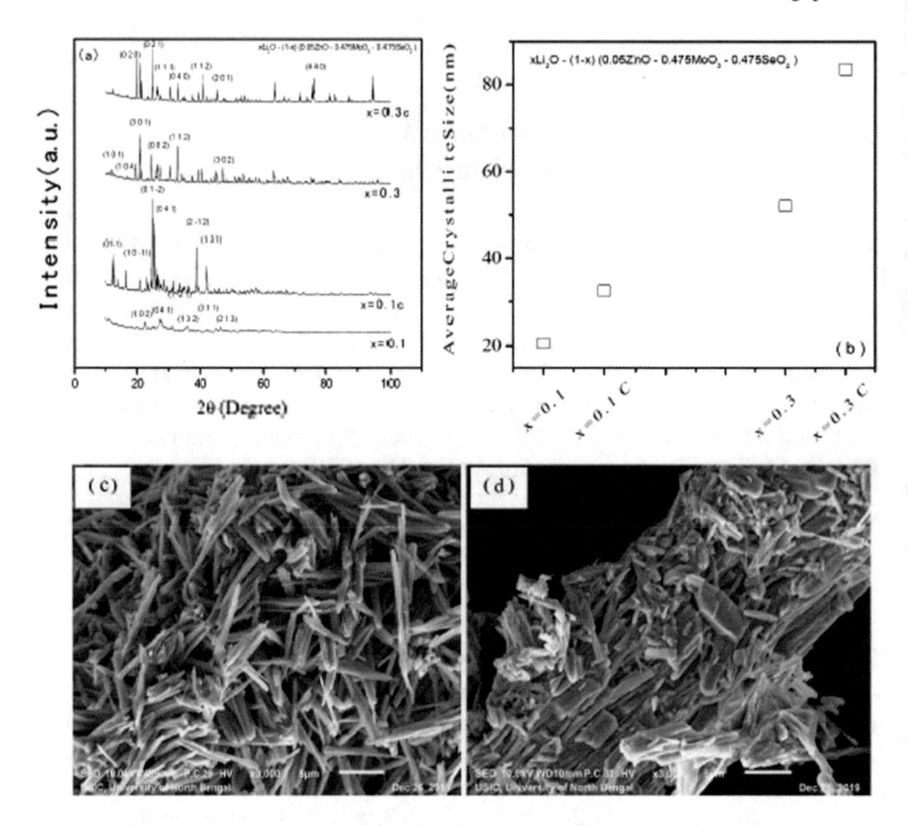

Fig. 12.1 **a** XRD of glassy system and crystalline counterparts; **b** average crystallite sizes with composition; **c** SEM micrographs for glassy system; **d** crystalline counterpart for $x = 0.1$

diameters, which are essentially related to DC resistivity. Figure 12.2a also shows null grain boundary effect [7]. Higher diameter for $x = 0.1$ (crystalline) indicates larger DC resistivity than that for $x = 0.1$ (glassy). Small arc in the lower frequency region (spur) is associated with polarization effect [7]. Here, parallel RC equivalent circuit may be included as the schematic circuit model for proper understanding of Li^+ motion.

Arrhenius relation of DC conductivity is:

$$\sigma_{DC} = \sigma_0 e^{-\left(\frac{E_\sigma}{KT}\right)} \tag{12.3}$$

where E_σ is the activation energy, T is the absolute temperature and k is the Boltzmann constant. DC conductivity data has been analyzed using Eq. (12.3), which shows thermally activated nature. DC conductivity decreases in the crystalline counterparts due to trapping of Li^+ ions in the extended voids (defects). E_σ has been estimated from the best fitted straight line fits. Figure 12.2b projects DC conductivity and E_σ with composition, which shows gradual reduction in DC conductivity. This result

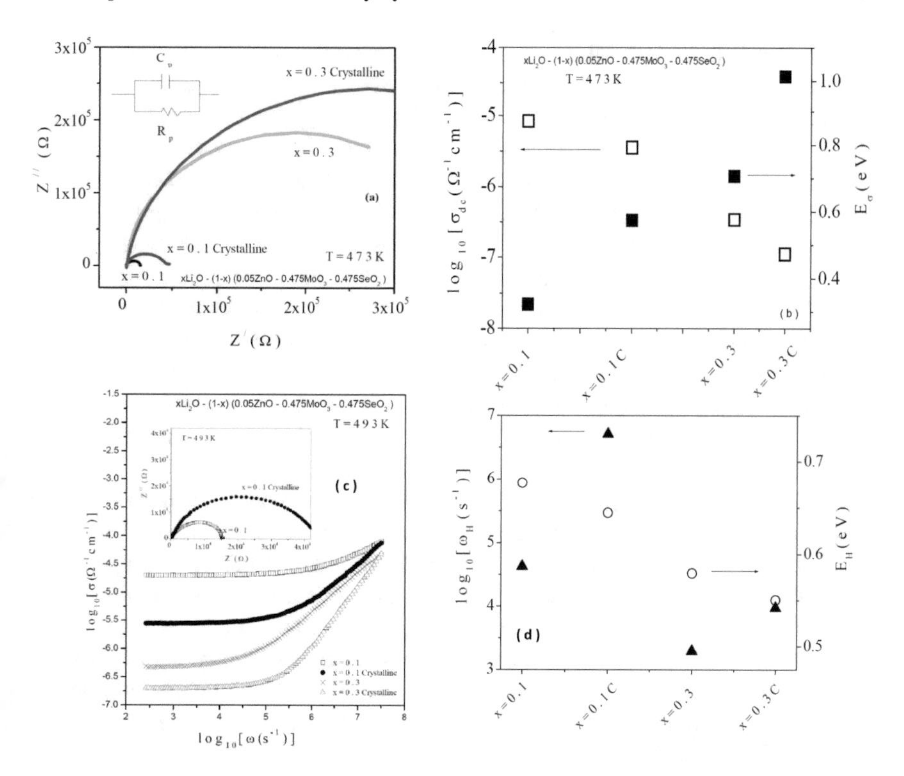

Fig. 12.2 **a** Complex impedance plots of glassy system and crystalline counterparts at 473 K. Equivalent circuit is in the inset; **b** DC conductivity and corresponding activation energy; **c** conductivity spectra at a temperature. Complex impedance plots and crystalline counterpart for $x = 0.1$ are included in the inset; **d** activation energy for hopping frequency and fixed temperature hopping frequency with compositions

suggests that larger sized nanophases are developed in the voids, where Li^+ ions are also expected to be trapped.

AC conductivity spectra are presented at a temperature in Fig. 12.2c, which can be analyzed [5–7] using Eq. (12.2). It is found to decrease with Li_2O content in a similar manner of DC conductivity. Complex impedance plots for $x = 0.1$ are included in the inset. It is noted that flat conductivity at low frequency is similar to the DC conductivity. This may lead to the diffusion motion of Li^+ ions [7]. At higher frequencies, AC conductivity shows dispersion and follows a power law as presented in Eq. (12.1). This may correspond to a nonrandom, correlated and subdiffusive interionic interaction [7] of Li^+ ions with an outcome of uniform change in power law exponent.

Understanding the conductivity process in Fig. 12.2c, attempts [8] have been made with a proposition of playing key role in the conduction of two types of frequency dependency. Firstly, the jump rate for hopping of charge carriers increases with applied frequency (except at very high temperatures) as per Eq. (94) and curve b of

Fig. 7 in Ref. [8]. Secondly, the carrier conductivity confined within spatial regions by disorder goes up with applied frequency [8]. The second effect [8] has to develop the very low-temperature (<10 K) AC conductivity of charge carriers that hop between especially close pairs of impurity states of very lightly doped compensated systems [9]. The prime issue is to interpret of such an effect, which should dominate the high-temperature hopping of high densities of charge carriers in the present glassy system, particularly for $x = 0.1$ and 0.3 glassy system. XRD and SEM studies reveal that various nanorods of definite dimensions are formed and present system has a tendency to break the $ZnSeO_3$ chain structure as the Li_2O content increases or to form crystalline counterparts by controlled heating. Consequently, more voids (defects) are expected to be formed. Since lithium ions are trapped in the voids near the high frequency and high-temperature region, frequency-dependent portion of the conductivity above the frequency-independent portion of the conductivity decreases with temperature.

Hopping frequency (ω_H) and a fractional power law exponent (n) have been estimated from the fitting of experimental conductivity spectra using Eq. (12.2). It is found that frequency exponent n is less than unity and depends significantly on composition due to some differences in structure. Scientific woks [7] show that n values are the indicators of the dimensionality of conduction pathways. $n \sim 2/3$ [7] may anticipate composition as well as temperature independency nature of three-dimensional motion of Li^+ ions. Two-dimensional ion motion in Na β-alumina and one-dimensional ion motion in hollandite [7] indicate smaller exponents, $n = 0.58 \pm 0.05$ and $n = 0.3 \pm 0.1$. So, AC conductivity is expected to be influenced by the dimensionality of the ion conduction space, which is manifested by the frequency exponent [6, 7]. Papathanassiou et al. [10] successfully developed a model based on the aspect of the distribution of the length of conduction paths regarding the universal power law dispersive conductivity in case of polymer networks and, generally, in disordered matter. Larger values of power law exponent (>1) may indicate percolation of charge carriers within limited frequency range [7].

Temperature dependence of ω_H exhibits thermally activated nature [7]. Fixed temperature hopping frequency (at 473 K) and corresponding activation energy are presented in Fig. 12.2d, which shows similar nature of DC conductivity and corresponding activation energy.

12.4 Advantages and Disadvantages

Some issues regarding advantages and disadvantage may be addressed for employing them in various cases. They are summarized as follows:

(a) Amorphous glassy systems:

   i. The conductivity of amorphous glassy systems is found to be higher than the heat-treated counterparts due to less disorders and grain boundaries.

ii. These systems show flexibility in wide range due to its structural and electrical properties.
iii. The complexity understanding is easier with respect to its crystalline counterparts.
iv. The main disadvantages of these glassy ceramics are their brittle nature which causes complexity in experimental setups.

(b) Heat-treated counterparts:

i. These systems exhibit lower conductivity due to structural defects as well as more grain boundaries.
ii. Due to complexity in understanding, it is having less use than the amorphous glassy systems.
iii. The advantage of these systems is its structural strength with make them less brittle and causes experimental ease.

12.5 Conclusion

Li_2O-doped glass-nanocomposites and their crystalline counterparts have been developed. Formation of $Li_2Zn_2(MoO_4)_3$, $ZnMoO_4$, $Zn(MoO_2)_2$, $Li_2Mo_6O_7$ and Li_2MoO_3 nanorods has been confirmed from XRD and SEM studies. Jonscher's universal power law and Almond-West formalism have been employed to interpret electrical conductivity data. Low-frequency conductivity and high-frequency dispersion have been well explained on the light of a nonrandom, correlated and sub-diffusive motion of Li^+. $ZnSeO_3$ chain structure has been formed, and molybdate rod-like structures are found to be interpenetrating into them. More voids should be responsible to trap Li^+.

Acknowledgements The financial assistance for the work by the Council of Scientific and Industrial Research (CSIR), India **[03 (1411)/ 17 / EMR-II],** is thankfully acknowledged.

References

1. K. Takada, J. Power Sources **394**, 74 (2018)
2. X. Yao, B. Huang, J. Yin, G. Peng, Z. Huang, C. Gao, D. Liu, X. Xu, Chin. Phys. B **25**, 018802 (2016)
3. K. Takada, T. Ohno, N. Ohta, T. Ohnishi, Y. Tanaka, ACS Energy Lett. **3**, 98 (2018)
4. A. Palui, A. Ghosh, Solid State Ionics **343**, 115126 (2019)
5. D.P. Almond, G.K. Duncan, A.R. West, Solid State Ionics **8**, 159 (1983)
6. G. Nanocomposites, *Synthesis* (Elsevier, Properties and Applications, 2016)
7. E.R. Shaaban, M.A. Kaid, M.G.S. Ali, J. Alloy. Compd. **613**, 324 (2014)
8. D. Emin, Adv. Phys. **24**, 305 (1975)
9. M. Pollak, T.H. Geballe, Phys. Rev. **122**, 1742 (1961)
10. A. N. Papathanassiou, I. Sakellis and J. Grammatikakis, Applied Phy. Lett. **91** (2007) 122911.

Part III
Applications of Li-doped Glass Composites

Chapter 13
Electrodes

Asmita Poddar, Madhab Roy, and Sanjib Bhattacharya

Abstract As efficient energy storage devices, batteries, including nickel–metal hydride (Ni-MH) batteries, lead acid batteries and lithium-ion batteries (LIBs) can be effectively combined with renewable energy sources such as solar energy, wind energy and hydrogen energy, and such batteries are expected to be advanced energy storage systems and reduce fossil fuel dependence. The lithium metal, used as a negative material in all solid-state battery systems, results in higher-energy density than what is currently available. Replacing liquid electrolytes with solid-state electrolytes (SSE) can also effectively inhibit the generation of SEI films and improve the cycle performance of batteries. Along with that, the size of the battery is reduced, and its application scope is expanded. Compared to liquid electrolytes, the most important property is that SSEs (e.g. inorganic ceramic electrolytes) do not leak and are non-flammable. Thus, safety is significantly boosted. The use of solid-state electrolytes can not only overcome the liquid electrolyte durability problem, but also provide an important path for developing next-generation LIBs. Research areas for lithium-ion batteries include extending lifetime, increasing energy density, improving safety, reducing cost and increasing charging speed, among others.

13.1 Introduction

Electrode, electric conductor, usually metal, used as either of the two terminals of an electrically conducting medium. It conducts current into and out of the medium, which may be an electrolytic solution as in a storage battery, or a solid, gas, or

---

A. Poddar
Department of Electrical Engineering, Dream Institute of Technology, Kolkata 700104, West Bengal, India

A. Poddar · M. Roy
Department of Electrical Engineering, Jadavpur University, Kolkata, West Bengal 700032, India

S. Bhattacharya (✉)
UGC-HRDC (Physics), University of North Bengal, Darjeeling 734013, West Bengal, India
e-mail: sanjib_ssp@yahoo.co.in; ddirhrdc@nbu.ac.in

S. Bhattacharya and K. Bhattacharya (eds.), *Lithium Ion Glassy Electrolytes*,
https://doi.org/10.1007/978-981-19-3269-4_13

vacuum. The electrode from which electrons emerge is called the cathode and is designated as negative; the electrode that receives electrons is called the anode and is designated as positive [1]. Nowadays, rapid technology development is taking place in the fields of portable electronic devices or electric cars which requires novel, technologically mature, safe and economically affordable portable power sources [2]. One of the factors that has crucial influence on batteries' performance is the electrical conductivity of their cathode materials. Therefore, it is very important to carry out researches on better cathode materials as well as methods to improve their properties [3]. In this context, the demand for more efficient and lighter Li-ion cells is growing extremely quickly.

13.2 Advantages of Li-Doped Glass Composites as Electrodes

A lithium-ion battery or Li-ion battery is a type of rechargeable battery where lithium ions move from the negative electrode through an electrolyte to the positive electrode during discharge and back when charging. Li-ion batteries use an intercalated lithium compound as the material at the positive electrode and typically graphite at the negative electrode.

Lithium-ion batteries are commonly used for portable electronics and electric vehicles and are growing in popularity for military and aerospace applications [4]. Nowadays, rapid technology development in the fields of portable electronic devices or electric cars requires novel, technologically mature, safe and economically affordable portable power sources. Especially, the demand for more efficient and lighter Li-ion cells is growing extremely quickly.

One of the factors that has crucial influence on batteries' performance is the electrical conductivity of their cathode materials. Therefore, it is very important to carry out researches on better cathode materials as well as methods to improve their properties.

British Chemist M. Stanley Whittingham first proposed the usage of lithium batteries, and he used titanium (IV) sulphide and lithium metal as the electrodes in his experiment [5]. But, this rechargeable lithium battery could never be made practical as titanium disulphide, even being synthesized under completely sealed conditions, when exposed to air, reacts to form hydrogen sulphide compounds, which have an unpleasant odour and are toxic to most animals.

For this, and other reasons, the development of Whittingham's lithium–titanium disulphide battery has been discontinued. The batteries with metallic lithium electrodes presented safety issues, as lithium metal reacts with water, releasing flammable hydrogen gas [6]. Consequently, research moved to development of batteries in which, instead of metallic lithium, only lithium compounds are present, being capable of accepting and releasing lithium ions.

13.3 Review Works on Li-Doped Glass Composites as Electrodes

In the year of 1973, Adam Heller proposed the lithium thionyl chloride battery [7], which are still used in implanted medical devices and in defence systems where a greater than 20-year shelf life, high-energy density and/or tolerance for extreme operating temperatures are required.

Basu, in 1977, demonstrated electrochemical intercalation of lithium in graphite at the University of Pennsylvania [8, 9]. This led to the remarkable development of a workable lithium intercalated graphite electrode at Bell Labs (LiC_6) [10] to provide an alternative to the lithium metal electrode battery.

Godshall et al. [11–13] and, shortly thereafter, Goodenough (Oxford University) and Koichi Mizushima (Tokyo University), worked in separate groups and demonstrated a rechargeable lithium cell with voltage in the 4 V range using lithium cobalt dioxide ($LiCoO_2$) as the positive electrode and lithium metal as the negative electrode [14, 15]. This innovation provided the positive electrode material that enabled early commercial lithium batteries.

$LiCoO_2$ is a stable positive electrode material which acts as a donor of lithium ions and thus can be used with a negative electrode material other than lithium metal [16]. By enabling the use of stable and easy-to-handle negative electrode materials, $LiCoO_2$ enabled novel rechargeable battery systems. Godshall et al. further identified few similar value of ternary compound lithium transition metal oxides such as the spinel $LiMn_2O_4$, Li_2MnO_3, $LiMnO_2$, $LiFeO_2$, $LiFe_5O_8$ and $LiFe_5O_4$ (and later lithium copper oxide and lithium nickel oxide cathode materials in 1985) [17].

In the year of 1980, Yazami validated the reversible electrochemical intercalation of lithium in graphite [18, 19] and invented the lithium graphite electrode (anode) [20, 21]. The organic electrolytes would decompose during charging with a graphite negative electrode. Yazami used a solid electrolyte to demonstrate that lithium could be reversibly intercalated in graphite through an electrochemical mechanism. As of 2011, Yazami's graphite electrode was the most commonly used electrode in commercial lithium-ion batteries.

The negative electrode has its origins in polyacenic semiconductive material (PAS) discovered by Yamabe and later by Yata in the early 1980s [22–25]. The seed of this technology was the discovery of conductive polymers by Professor Hideki Shirakawa and his group, and it could also be seen as having started from the polyacetylene lithium-ion battery developed by Alan MacDiarmid and Alan J. Heeger et al. [26]

In 1982, Godshall et al. were awarded US Patent 4,340,652 [27] for the use of $LiCoO_2$ as cathodes in lithium batteries, based on Godshall's Stanford University Ph.D. dissertation and 1979 publications.

In1983, Michael M. Thackeray, Peter Bruce, William David and John B. Goodenough developed manganese spinel, Mn_2O_4, as a charged cathode material for lithium-ion batteries, and study has shown that two flat plateaus occur on discharge with lithium, one at 4 V, stoichiometry $LiMn_2O_4$ and one at 3 V with a final stoichiometry of $Li_2Mn_2O_4$ [28].

Akira Yoshino assembled a prototype cell using carbonaceous material in 1985, into which lithium ions could be inserted as one electrode and lithium cobalt oxide (LiCoO) as the other[29]. This histrionically improved safety. LiCoO facilitated industrial-scale production and initiated the usage of commercial lithium-ion battery.

In the year of 1989, Manthiram and Goodenough discovered the polyanion class of cathodes [30, 31]. Their study has shown that positive electrodes containing polyanions, e.g. sulphates, produce higher voltages than oxides due to the inductive effect of the polyanion. This polyanion class contains materials such as lithium iron phosphate [32].

From 1990, lithium-ion batteries are produced for commercialization, and advancements have been made for better performance and increased capacity of lithium-ion batteries. In the year of 1991, Sony and Asahi Kasei released the first commercial lithium-ion battery [33]. The Japanese team that successfully commercialized the technology was led by Yoshio Nishi. In 1996 [34], Goodenough, Padhi and co-workers proposed lithium iron phosphate ($LiFePO_4$) and other phospho-olivines (lithium metal phosphates with the same structure as mineral olivine) as positive electrode materials [35].

Johnson et al. reported the discovery of the high capacity high-voltage lithium-rich NMC cathode materials in 1998 [36].

In 2001, Manthiram and co-workers discovered that the capacity limitations of layered oxide cathodes are a result of chemical instability that originates due to the relative positions of the metal 3*d* band relative to the top of the oxygen 2*p* band [37–39]. This discovery has had significant implications for the practically accessible compositional space of lithium-ion battery layered oxide cathodes, as well as their stability from a safety perspective.

In the same year, Christopher, Thackeray et al., filed a patent [40, 41] for lithium nickel manganese cobalt oxide (NMC) lithium-rich cathodes based on a domain structure, whereas Lu and Dahn filed a patent [42] for the NMC class of positive electrode materials, which offers safety and energy density improvements over the widely used lithium cobalt oxide.

In the year of 2002, Yet-Ming Chiang and his group at MIT showed a substantial improvement in the performance of lithium batteries by boosting the material's conductivity by doping it [43] with aluminium, niobium and zirconium. The exact mechanism causing the improvement became the subject of widespread debate [44].

In 2004, Yet-Ming Chiang again increased performance by utilizing lithium iron phosphate particles of less than 100 nanometres in diameter. This method decreased particle density almost one 100-fold, increased the positive electrode's surface area and improved capacity and performance. Commercialization led to a rapid growth in the market for higher capacity lithium-ion batteries, as well as a patent infringement battle between Chiang and John Goodenough [44].

In 2005, Song et al. reported a new two-electron vanadium phosphate cathode material with high-energy density [45, 46].

In 2011, lithium nickel manganese cobalt oxide (NMC) cathodes, developed at Argonne National Laboratory, are manufactured commercially by BASF in Ohio

[47]. In the same year, lithium-ion batteries accounted for 66% of all portable secondary (i.e. rechargeable) battery sales in Japan [48].

In the year of 2012, Goodenough, Yazami and Yoshino received the 2012 IEEE Medal for Environmental and Safety Technologies for developing the lithium-ion battery [21].

Goodenough, Nishi, Yazami and Yoshino were awarded the Charles Stark Draper Prize of the National Academy of Engineering for their pioneering efforts in the field in 2014 [49]. In the same year, commercial batteries from Amprius Corp. reached 650 Wh/L (a 20% increase), using a silicon anode, and were delivered to customers [50].

In 2016, Koichi Mizushima and Akira Yoshino received the NIMS Award from the National Institute for Materials Science, for Mizushima's discovery of the $LiCoO_2$ cathode material for the lithium-ion battery and Yoshino's development of the lithium-ion battery [51]. In the same year, Qi and Koenig reported a scalable method to produce sub-micrometre-sized $LiCoO_2$ using a template-based approach [52]. In 2010, global lithium-ion battery production capacity was 20 gigawatt-hours [53]. By 2016, it reached 28 GWh.

13.4 Materials Acceptable for Application Parameters

The study on effects of lithium oxide on the thermal properties of the ternary system glasses (Li_2O_3–B_2O_3–Al_2O_3) has been carried out [54]. The boric anhydride B_2O_3 is used as the only trainer of network. It introduces many valuable properties on the glass system like improving the fusibility, increasing the mechanical resistance, high thermal resistance and a decrease in the surface tension and increases the chemical resistances [55, 56]. The study on physical properties show that the density of the samples increases with the addition of lithium oxide, which can be explained by filling the voids between the structural units by the network modifiers (Ion of Li^+) [54]. Thus, the molar volume decreases and density increases. Lithium has applications in the field of optometry too. The study on optical properties of lithium borate glass $(Li_2O)_x$–$(B_2O_3)_{1-x}$ shows that the refractive index increases with decreasing molar volume and which in turn increases the density [57]. A series of $(Li_2O)x$–$(B_2O_3)_{1-x}$has been synthesized, and the structure of the glass system was determined by FTIRand X-ray diffraction. An addition of Li_2O causes change in coordination number, and the coordination number of lithium borate glass leads to the increase of refractive index and creates more non-bridging oxygen which results in the decrease of optical energy band gap for both direct and indirect band gap. It is used for detecting penetration of radiation which is applied in homeland security and non-proliferation. The main objectives of the present work were to study the refractive index and optical band gap with variation of lithium borate glass composition. The characterization and properties of lithium disilicate glass ceramic ($Li_2Si_2O_5$) are studied in the SiO_2–Li_2O–K_2O–Al_2O_3 system [58]. The experimental results and their discussion show that it is found as one such all-ceramic system that is currently used in the fabrication of single and multiunit dental restorations. The area of applications is mainly for dental crowns, bridges and veneers because of its colour being similar to natural teeth and its excellent mechanical properties [59].

It is seen that unconventional bismuthate glassy samples containing lithium oxide have been prepared by a conventional melt quench technique [60]. The study of X-ray diffraction, scanning electron microscopy and differential thermal analysis shows that stable binary glasses of composition $xLi_2O–(100–x)Bi_2O_3$ can be achieved for $x = 20 - 35$ mol%. Differential thermal analysis and optical studies show that the strength of the glass network decreases with the increase of Li_2O content in the glass matrix. A study on synthesis, thermal, structural and electrical properties of vanadium-doped lithium manganese borate glass and nanocomposites has been done [61]. A glassy sample with a nominal formula $LiMn_{1-3x/2}–VxBO_3$ (where $x = 0.05$) was synthesized, and dependencies of glass transition and crystallization temperatures on the heating rate in DTA experiments were determined. It is found that the highest conductivity at room temperature (2.6×10^{-9} Scm^{-1}) was obtained for the sample nano-crystallized at 700 °C. This means conductivity increases by a factor of 2×10^6, which was also higher by three orders of magnitude than for lithium manganese borate glass without vanadium.

The glass system $Li_2O–B_2O_3–ZnO$ (LBZ) prepared by melt quenched method has been studied [62]. The effect of adding alkali oxide Li_2O on the variation of the structure and properties of the glass system ($B_2O_3–ZnO$) led to interesting results. It is noted that the density of glasses decreases with the decrease of lithium oxide and increase of boron oxide. The addition of the network modifying oxide Li_2O content greater than 50% (by weight in %) in a system glass ($B_2O_3–ZnO$) contributes to the creation of the non-bridging oxygen and thus the reduction in the structural rigidity of vitreous network and weakening of the mechanical and thermal properties.

Lithium has applications in food industry too. The lithium lanthanum titanate ceramics as sensitive material for pH-sensors have been carried out [63]. The high lithium conductivity of this oxide, at room temperature, would indicate a possible use of this material as a lithium-ion-selective electrode. This sensor can be used in industrial processes like milk fermentation, in control of cleaning of fermenters (cleaning in place), yoghurt fabrication and wastewater treatments.

It has been studied that Co_3O_4 nanotubes, nano-rods and nanoparticles are used as the anode materials of lithium-ion batteries [64]. The results show that the Co_3O_4 nanotubes, prepared by a porous alumina template method, display high discharge capacity and superior cycling reversibility.

Studies show that the nanostructured silicon is promising for high capacity anodes in lithium batteries [65]. The specific capacity of silicon is an order of magnitude higher than that of conventional graphite anodes, but the large volume change of silicon during lithiation and delithiation and the resulting poor cyclability has prevented its commercial application.

Frequency-dependent AC conductivity measurements were implied on the prepared silver-doped lithium tellurite borate glass system in the high-frequency range 5 Hz–35 MHz [66]. Lithium containing glasses show higher-order electrical conductivity or ionic conductivity and are widely used for electro-chemical devices like solid-state batteries, glass electrolytes and fuel cells. The four most predominant compounds formed from lithium reactions are lithium hydride (LiH), lithium oxide (Li_20), lithium nitride (Li_3N) and lithium hydroxide (LiOH). All are stable but extremely reactive and corrosive compounds [67–70]. No metal or refractory material can handle molten lithium hydroxide (LiOH) in high concentrations as LiOH, being enough corrosive. Li_20 is highly reactive with water, carbon dioxide and refractory compounds. Li_3N is also very reactive. No metal or ceramic has been found resistant to molten nitride. Being hygroscopic, it forms ammonia in the presence of water, whereas LiH reduces oxides, chlorides, sulphides readily and reacts with metals and ceramics at high temperatures.

Apart from these applications, Li_2O containing glasses are also used for various industrial applications as well as in the formation of dielectric materials for high-speed transmission of signals.

13.5 Conclusion

Rechargeable batteries play a crucial role in portable electronics, transportation, backup power and load-levelling applications. Presently, lead acid and Li-ion chemistries are the most important categories of rechargeable batteries. The Li-ion chemistry is expected to play a superior role in the future due to its greater gravimetric and volumetric densities as compared to other battery chemistries. Moreover, the cost of Li-ion batteries continues to drop due to engineering progresses and materials developments. The nanocomposite materials are at the epicentre of the material advancements in cathode, anode, binder and separator of Li-ion batteries. The great advantage of nanocomposite materials is that they can improve the safety, cycle life, rate capability and specific capability of Li-ion batteries.

References

1. E. Britannica, The Editors of Encyclopaedia Britannica. Electrode (2011)
2. J.A. Sanguesa, V. Torres-Sanz, P. Garrido, F.J. Martinez, J.M. Marquez barja, A Review on electric vehicles: technologies and challenges. Smart Cities **4**, 372–404 (2021). https://doi.org/10.3390/smartcities4010022
3. N. Mohamed, N.K. Allam, Recent advances in the design of cathode materials for Li-ion batteries. Rev. RSC Adv. (2020)
4. M.S. Ballon, Electrovaya, Tata motors to make electric Indica cleantech.com. Archived from the original, on 9 May 2011. Retrieved 11 June 2010 (2008)
5. Binghamton Professor recognized for energy research. The Research Foundation for the state university of New York. Retrieved 10 October 2019
6. XXIV.—On chemical analysis by spectrum-observations. Q. J. Chem. Soc. Lond. **13**(3), 270 (1861). https://doi.org/10.1039/QJ8611300270
7. A. Heller, Electrochemical Cell United States Patent. Retrieved 18 Nov 2013

8. M. Zanini, S. Basu, J.E. Fischer, Alternate synthesis and reflectivity spectrum of stage 1 lithium—graphite intercalation compound. Carbon **16**(3), 211–212 (1978). https://doi.org/10.1016/0008-6223(78)90026-X
9. S. Basu, C. Zeller, P.J. Flanders, C.D. Fuerst, W.D. Johnson, J.E. Fischer, Synthesis and properties of lithium-graphite intercalation compounds. Mater. Sci. Eng. **38**(3), 275–283 (1979). https://doi.org/10.1016/0025-5416(79)90132-0
10. US 4304825, Basu, issued 8 December 1981, Rechargeable battery, assigned to Bell Telephone Laboratories
11. N.A. Godshall, I.D. Raistrick, R.A. Huggins, Thermodynamic investigations of ternary lithium-transition metal-oxygen cathode materials. Mater. Res. Bull. **15**(5), 561 (1980). https://doi.org/10.1016/0025-5408(80)90135-X
12. N.A. Godshall, Electrochemical and thermodynamic investigation of ternary lithium-transition metal-oxide cathode materials for lithium batteries: Li_2MnO_4spinel, $LiCoO_2$, and $LiFeO_2$, Presentation at 156th Meeting of the Electrochemical Society, Los Angeles, CA (1979)
13. N.A. Godshall, Electrochemical and thermodynamic investigation of ternary lithium-transition metal-oxygen cathode materials for lithium batteries. Stanford University. ProQuest Dissertations Publishing, 1980. 8024663 (1980)
14. J. Goodenough, USPTO search for inventions. Patft.uspto.gov. Retrieved 8 October 2011
15. K. Mizushima, P.C. Jones, P.J. Wiseman, J.B. Goodenough, Li_xCoO_2(0<x<-1): A new cathode material for batteries of high energy density. Mater. Res. Bull. **15**(6), 783–789 (1980). https://doi.org/10.1016/0025-5408(80)90012-4
16. P. Poizot, S. Laruelle, S. Grugeon, J. Tarascon, Nano-sized transition-metal oxides as negative-electrode materials for lithium-ion batteries. Nature **407**(6803), 496–499 (2000). https://doi.org/10.1038/35035045
17. N. Godshall, Lithium transport in ternary lithium-copper-oxygen cathode materials. Solid State Ionics **18–19**, 788–793 (1986). https://doi.org/10.1016/0167-2738(86)90263-8
18. International Meeting on Lithium Batteries, Rome, 27–29 April 1982, C.L.U.P. Ed. Milan, Abstract #23
19. R. Yazami, P. Touzain, J. Power Sources **9**(3), 365–371 (1983). https://doi.org/10.1016/0378-7753(83)87040-2
20. R. Yazami, National Academy of Engineering. Retrieved 12 Oct 2019
21. IEEE Medal for Environmental and Safety Technologies Recipients. IEEE Medal for Environmental and Safety Technologies. Institute of Electrical and Electronics Engineers. Retrieved 29 July 2019
22. T. Yamabe, Lichiumu Ion NijiDenchi: KenkyuKaihatu No GenryuWoKataru Lithium ion rechargeable batteries: tracing the origins of research and development: focus on the history of negative-electrode material development. J. Kagaku (in Japanese). **70**(12), 40–46 (2015). Archived from the original on 8 August 2016. Retrieved 15 June 2016
23. P. Novák, K. Muller, K.S.V. Santhanam, O. Haas, Electrochemically active polymers for rechargeable batteries. Chem. Rev. **97**(1), 271–272 (1997). https://doi.org/10.1021/cr941181o. PMID11848869
24. T. Yamabe, K. Tanaka, K. Ohzeki, S. Yata, Electronic structure of polyacenacene. a one-dimensional graphite. Solid State Commun. **44**(6), 823 (1982). https://doi.org/10.1016/0038-1098(82)90282-4
25. US 4601849, S. Yata, Electrically conductive organic polymeric material and process for production thereof
26. P.J. Nigrey, Lightweight rechargeable storage batteries using polyacetylene $(CH)_x$ as the cathode-active material. J. Electrochem. Soc. **128**(8), 1651 (1981). https://doi.org/10.1149/1.2127704
27. N.A. Godshall, I.D. Raistrick, R.A. Huggins, U.S. Patent 4,340,652, Ternary compound electrode for lithium cells, issued 20 July 1982, filed by Stanford University on 30 July 1980
28. M.M. Thackeray, W.I.F. David, P.G. Bruce, J.B. Goodenough, Lithium insertion into manganese spinels. Mater. Res. Bull. **18**(4), 461–472 (1983). https://doi.org/10.1016/0025-5408(83)90138-1

29. US 4668595, A. Yoshino, Secondary Battery, issued 10 May 1985, assigned to Asahi Kasei
30. A. Manthiram, J.B. Goodenough, Lithium insertion into $Fe_2(SO_4)_3$ frameworks. J. Power Sources **26**(3–4), 403–408 (1989). https://doi.org/10.1016/0378-7753(89)80153-3
31. A. Manthiram, J.B. Goodenough, Lithium insertion into $Fe_2(MO_4)_3$ frameworks: comparison of M = W with M = Mo. J. Solid State Chem. **71**(2), 349–360 (1987). https://doi.org/10.1016/0022-4596(87)90242-8
32. C. Masquelier, L. Croguennec, Polyanionic (phosphates, silicates, sulfates) frameworks as electrode materials for rechargeable Li (or Na) batteries. Chem. Rev. **113**(8), 6552–6591 (2013). https://doi.org/10.1021/cr3001862.PMID23742145
33. Keywords to understanding Sony Energy Devices—keyword 1991. Archived from the original on 4 Mar 2016
34. Y. Nishi, National Academy of Engineering. Retrieved 12 Oct 2019
35. A.K. Padhi, K.S. Naujundaswamy, J.B. Goodenough, $LiFePO_4$: a novel cathode material for rechargeable batteries. Electrochem. Soc. Meet. Abs. **96–1**, 73 (1996)
36. C.S. Johnson, J.T. Vaughey, M.M. Thackeray, T.E. Bofinger, S.A. Hackney, Layered lithium-manganese oxide electrodes derived from rock-salt $Li_xMn_yO_z$ ($x + y = z$) precursors, in *194th Meeting of the Electrochemical Society* (Boston, MA, 1998)
37. R.V. Chebiam, A.M. Kannan, F. Prado, A. Manthiram, Comparison of the chemical stability of the high energy density cathodes of lithium-ion batteries. Electrochem. Commun. **3**(11), 624–627 (2001). https://doi.org/10.1016/S1388-2481(01)00232-6
38. R.V. Chebiam, F. Prado, A. Manthiram, Soft chemistry synthesis and characterization of layered $Li_{1-x}Ni_{1-y}Co_yO_{2-\delta}$ ($0 \leq x \leq 1$ and $0 \leq y \leq 1$). Chem. Mater. **13**(9), 2951–2957 (2001). https://doi.org/10.1021/cm0102537
39. A. Manthiram, A reflection on lithium-ion battery cathode chemistry. Nature Commun. **11**(1), 1550 (2020). https://doi.org/10.1038/s41467-020-15355-0
40. US US6677082, M. Thackeray, K. Amine, J.S. Kim, Lithium metal oxide electrodes for lithium cells and batteries
41. US US6680143, M. Thackeray, K. Amine, J.S. Kim, Lithium metal oxide electrodes for lithium cells and batteries
42. US US6964828 B2, Z. Lu, Cathode compositions for lithium-ion batteries
43. S.Y. Chung, J.T. Bloking, Y.M. Chiang, Electronically conductive phospho-olivines as lithium storage electrodes. Nat. Mater. **1**(2), 123–128 (2002). https://doi.org/10.1038/nmat732
44. In search of the perfect battery. The Economist. 6 March 2008. Archived from the original (PDF) on 27 July 2011. Retrieved 11 May 2010
45. Y. Song, P.Y. Zavalij, M.S. Whittingham, ε-VOPO4: electrochemical synthesis and enhanced cathode behavior. J. Electrochem. Soc. **152**(4), A721–A728 (2005). https://doi.org/10.1149/1.1862265
46. S.C. Lim, J.T. Vaughey, W.T.A. Harrison, L.L. Dussack, A.J. Jacobson, J.W. Johnson, Redox transformations of simple vanadium phosphates: the synthesis of ϵ-VOPO4. Solid State Ionics **84**(3–4), 219–226 (1996). https://doi.org/10.1016/0167-2738(96)00007-0
47. BASF breaks ground for lithium-ion battery materials plant in Ohio, October 2009
48. Monthly battery sales statistics. Machinery statistics released by the Ministry of Economy, Trade and Industry, March 2011
49. Lithium Ion Battery Pioneers Receive Draper Prize, Engineering's Top Honor Archived 3 April 2015 at the Wayback Machine, University of Texas, 6 January 2014
50. At long last, new lithium battery tech actually arrives on the market (and might already be in your smartphone). ExtremeTech. Retrieved 16 February 2014
51. NIMS Award Goes to Koichi Mizushima and Akira Yoshino. National Institute for Materials Science. 14 September 2016. Retrieved 9 Apr 2020
52. Z. Qi, G.M. Koenig, High-performance $LiCoO_2$ sub-micrometer materials from scalable microparticle template processing. Chem Select. **1**(13), 3992–3999 (2016). https://doi.org/10.1002/slct.201600872
53. Lithium-ion batteries for mobility and stationary storage applications (PDF). European Commission. Archived (PDF) from the original on 14 July 2019. global lithium-ion battery production from about 20GWh in 2010

54. D. Aboutaleb, B. Safi, Lithium oxide effect on the thermal and physical properties of the ternary system glasses (Li_2O_3-B_2O_3-Al_2O_3). World Acad. Sci. Eng. Technol. Int. J. Chem. Mol. Nucl. Mater. Metall. Eng. **9**(3) (2015)
55. J.E. Shelby, Introduction to glass, science and technologies, in *Immiscibility/Phase Separation* (The Royal Society of Chemistry, 1997), pp. 48–67
56. J. Barton, Claude Guillem, Les verres: science et technologie, Chimie I matériaux, Edition Edp
57. M.K. Halimah, W.H. Chiew, H.A.A. Sidek, W.M. Daud, Z.A. Wahab, A.M. Khamirul, S.M. Iskandar, Optical properties of lithium borate glass $(Li_2O)x$ $(B_2O_3)_{1-x}$. Sains Malaysiana **43**(6), 899–902 (2014)
58. N. Monmaturapoj, P. Lawita, W. Thepsuwan, Characterisation and properties of lithium disilicate glass ceramics in the SiO_2–Li_2O–K_2O–Al_2O_3 system for dental applications. National Metal and Materials Technology Center, 114 Thailand Science Park, Pathumthani 12120, Thailand. Received 29 April 2013; Revised 25 June 2013; Accepted 28 June 2013
59. E. El-Meliegy, R. van Noort, *Glasses and Glass Ceramics for Medical Applications* (Springer, New York, NY, USA, 2012)
60. S. Hazra, S. Mandal, A. Ghosh, Properties of unconventional lithium bismuthate glasses, in *Solid State Physics Department, Indian Association for the Cultivation of Science*, Calcutta-700 032, India. Received 9 May 1997
61. A. Jarocka, P.P. Michalski, J. Ryl, M. Wasiucionek, J.E. Garbarczyk, T.K. Pietrzak, Synthesis, thermal, structural and electrical properties of vanadium-doped lithium-manganese-borate glass and nano-composites. Received: 25 March 2019/Revised: 1 August 2019/Accepted: 23 August 2019 © The Author(s) 2019
62. D. Aboutaleb, B. Safi, S. Laichaoui, Z. Lemou, Effect of Li_2O and Na_2O addition on structure and properties of glass system (B_2O_3–ZnO). Int. J. Mater. Mech. Manuf. **6**(6) (2018)
63. C. Bohnke, H. Duroy, J.-L. Fourquet. pH sensors with lithium lanthanum titanate sensitive material: applications in food industry. Sens. Actuators B Chem. **89**(3), 240–247 (2003)
64. W.Y. Li, L.N. Xu, J. Chen, Co_3O_4 Nano-materials in lithium-ion batteries and gas sensors. Adv. Func. Mater. (2005). https://doi.org/10.1002/adfm.200400429
65. J.R. Szczech, S. Jin, Nanostructured silicon for high capacity lithium battery anodes. Received 19th July 2010, Accepted 21st October 2010. https://doi.org/10.1039/c0ee00281j
66. P. Naresh, V. Sunitha, A. Padmaja, P. Umal, N. Narsimlu, M. Srinivas, R. NP, K. Siva Kumar, Dielectric studies of silver doped lithium tellurite borate glasses for fast ionic battery applications, in *AIP Conference Proceedings*, vol. 2244, p. 100002 (2020). https://doi.org/10.1063/5.0008992
67. T.Y. Tlen, F.A. Hummel, Department of Ceramic Technology, The Pennsylvania State University, University Park, Pennsylvania
68. R.C. Weast (ed.), *Handbook of Chemistry and Physics*, 57th edn (CRC Press, 1976–77)
69. P.E. Landolt, M. Sittig, Lithium, in *Rare Metals Handbook*, 2nd edn (1961), pp. 239–252
70. J.H. Perry, *Chemical Engineer's Hand book*, 4th edn. (McGraw-Hill Book Company, NY, 1973)

Chapter 14
Photonic Glass Ceramics

Swarupa Ojha, Madhab Roy, and Sanjib Bhattacharya

Abstract This chapter deals with glass ceramics and their attracting properties which make them promising materials in various photonic applications like optical amplifiers, telecommunications, spectroscopy, solid-state lasers, light detectors, etc. A brief history on the development of ceramic glasses using different materials has been illustrated. A brief introduction about chalcogenide glasses along with their unique properties has been presented with the main focus on their optical properties for photonic applications. There is a small discussion on the materials, for making chalcogenide glasses, which have been accepted for various application parameters. Chalcogenide systems containing lithium ions are exceptionally promising candidates for various photonic applications.

Keywords Glass ceramics · Photonic glass ceramics · Chalcogenide glasses · Optical properties and applications of chalcogenide glasses

14.1 Photonic Glass Ceramics—What Is It?

Glass ceramics are nanomaterials in which crystals, in the size range of nanometers to micrometres, are embedded in a glass matrix. Glass ceramic is a polycrystalline material which can be formed by crystallizing base glass in controlled manner. Generally, crystallinity of these glasses lies between 30 and 70% [1]. This material exhibits properties of glasses (like simple fabrication, moulding, etc.) as well as ceramics (such as hardness, high thermal shock resistance). This material is having one amorphous

S. Ojha
Department of Electrical Engineering, OmDayal Group of Institutions, Howrah 711316, West Bengal, India

S. Ojha · M. Roy
Department of Electrical Engineering, Jadavpur University, Jadavpur, Kolkata 700032, India

S. Bhattacharya (✉)
UGC-HRDC (Physics), University of North Bengal, Darjeeling 734013, West Bengal, India
e-mail: sanjib_ssp@yahoo.co.in; ddirhrdc@nbu.ac.in

S. Bhattacharya and K. Bhattacharya (eds.), *Lithium Ion Glassy Electrolytes*,
https://doi.org/10.1007/978-981-19-3269-4_14

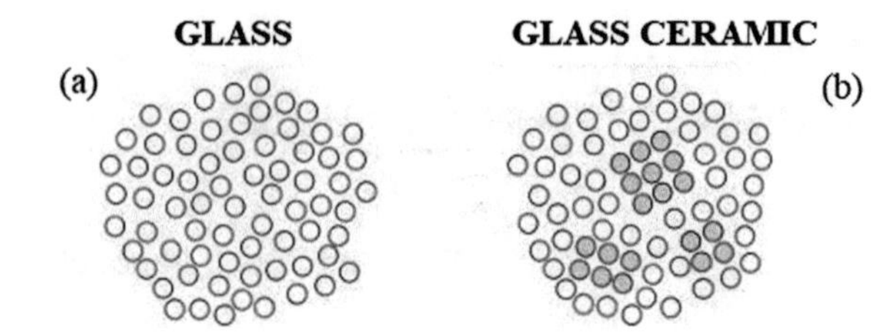

Fig. 14.1 Schematic of two-dimensional structures: **a** glass **b** glass ceramics

phase and one or more crystalline phases which enhance the need for controlled crystallization during its production [1]. This property of glass ceramic makes it different from other glasses in which spontaneous crystallization is required. Some of the fascinating properties of glass ceramics are zero void fraction (or porosity), high strength, hardiness, opacity, low thermal expansion (material's tendency of changing its shape and density with temperature), fluorescence (property of a material to emit light in absorbing electromagnetic radiation), high-temperature stability, high chemical durability, low dielectric constant and loss, etc. [1]. In a glass ceramic, it is possible to modify these properties by varying its glass compositions and crystallizing it in controlled mode [1]. Figure 14.1a, b shows the schematic of two-dimensional structures of glass and glass ceramics, respectively.

Glass ceramics are nanocomposite materials in which nanocrystals are embedded in a glass matrix. The properties of such materials are determined by their compositions and the volume fractions of crystalline and amorphous phases in the materials. For the materials having confined structures, transparency is a vital property.

The attracting properties of ceramic glasses have made these materials as one of the most promising materials in photonic (which is a technology to generate, control and detect photons across electromagnetic spectrum) applications like optical amplifiers, telecommunications, spectroscopy, solid-state lasers, light detectors, etc. [2]. In these glasses because of the presence of crystals in the size range of certain micrometres to nanometres and due to the small difference of refractive index between the crystalline and the amorphous phases, the high transparency is retained by the materials [2]. In glass ceramics, the amount of crystalline volume fraction may vary from 30 to 90%. They can also be used as luminescent materials as the addition of impurities plays a significant role in modification of transparency in glass ceramics [2].

14.2 Background

The glass ceramics were discovered by a renowned American scientist S. Donald Stookey, who was associated with Corning Inc., in the year 1952 [3]. He discovered the first glass ceramic from a glass material named 'Fotoform' due to its overheating at 900 °C. Instead of getting an expected melted mess, he obtained an opaque white plate which was found to be much stronger than the glass material Fotoform [4]. After that, in the year 1958, he had also discovered two more glass ceramics. The first one

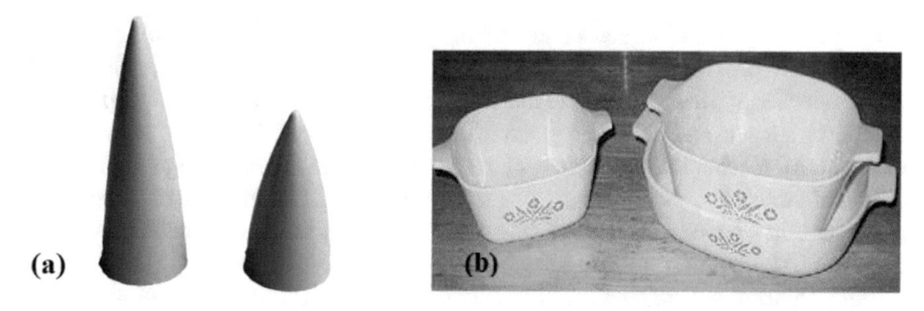

Fig. 14.2 Ceramic glasses as **a** Missile Radome and **b** CorningWare Kitchenware

was suitable as Radome in the nose core of missiles, and the second one was found as suited for consumer kitchenware. The glass ceramic which was found suitable in kitchenware was named as Pyroceram [5]. Some of the lucrative properties of Pyroceram were light in weight, high heat tolerance, durable, low thermal expansion, etc. The brand name of the products made from Pyroceram was CorningWare [5]. Ceramic glasses as Missile Radome and CorningWare Kitchenware are shown in Fig. 14.2a, b, respectively.

The control of nucleation and crystallization in the base glass is the key to form a glass ceramic. In 1962, Corning Inc. developed a glass ceramic named 'Chemcor' which was stronger than Pyroceram. 'Zerodur'—a transparent glass ceramic was developed by Schott AG in 1968. Its applications are found in very big telescope mirrors and as reference for testing the validity of new glass structural models [4]. Nippon Electric glass produced 'FireLite', which was also a transparent glass ceramic, in 1988 [6]. It was mainly used in safety products like fire rated doors. Some companies, like technical glass products, are manufacturing this glass ceramic even now [7]. Reports on development and studies on formation mechanisms of numerous ceramic glasses are available [8–17]. The study on properties, structures and applications of ceramic glasses have been reported by Holand and Beall [18].

With the advancement in the development of glass ceramics, their demand has been increased in various applications in many fields like health and medicines, energy fields (like nuclear industry for immobilizing waste and generation, transportation and storage of electricity, rechargeable batteries), transport industries (as valves, oxygen sensors, catalysts, etc.) and communication technologies (in motors, filters, etc.) [19]. With the more development of glass ceramics, their applications in the optical and photonic devices have also been broadening. SiO_2 glasses are the most available materials used in optical technology. The wide optical band gaps of ceramics glasses, which are insulating in nature, make them suitable for being used in optical devices [19]. The processing conditions of ceramic glasses can adapt transparency because their response to light propagation is strongly structure dependent. Glass ceramics are having many advantages over single crystals. Simple fabrication steps are required in the production of glass ceramics. It is easy to produce large pieces of ceramics having arbitrary shape and size, which is quite impossible to obtain in single crystals. In single crystal, high impurity concentration doping is not possible, while in glass ceramics, the possibility of impurity doping helps in getting many properties which make these glasses very unique and attractive [19]. The properties of these glasses are mainly controlled by compositions. As compared to amorphous glasses, ceramics glasses have shown many advantageous properties like superior mechanical resistance, better toughness and hardness, high thermal and chemical stabilities, etc. Because of the excellent thermal and chemical stability, glass ceramics have proved their suitability in a wide range of applications from cooktops to large telescope mirrors.

For preparing a ceramic glass melt, quenching method is generally used to form its glassy state followed by controlled heat treatment for partially crystallization. Due to no requirement of pressing or sintering during fabrication, these glasses contain no pores which make these glasses different from sintered ceramics. The first photonic application of glass ceramic was reported in 1995 by Tick et al. [20] in which utilization of oxyfluoride ceramic glass as optical amplifiers to be operated at 1300nm was mentioned. The suitability of photonics glass ceramics has also been found in waveguide applications like propagation of light and enhancement of luminescence in which transparency turns out to be vital [21]. In 1998, Tick et al. suggested some wide-ranging criteria for light propagation which includes crystallite size, particle distribution and clustering [22].

14.3 Chalcogenide Glassy Systems

The glasses which are made from chalcogens, i.e. group 16 elements of periodic table, are called chalcogenide glasses. These glasses are formed by combining sulphur, selenium and/or tellurium with some electropositive elements, like group 14 or group 15 elements, of periodic table. These glasses generally consist of covalent bonds. The percentage of ionic conductivity in these glasses is around 9%. These glasses are different from oxide glasses (like SiO_2) because of their diverse attractive properties. Chalcogenide glasses have proved their efficacy in many optical and photonic applications, spectral range of which is around 0.6–15 μm. These glasses, due to wide infrared transparency widow, have been found suitable in preparing many active devices (like laser fibre amplifiers and nonlinear components) as well as passive devices (such as windows, lenses) [23]. The reason of showing good transparency of these glasses, as compared to oxide glasses, in the mid-infrared spectrum is their low phonon energies [23]. These glasses exhibit extraordinary nonlinear properties because of the presence of polarizable atoms and lone electron pairs. Their linear refractive indices are very high, and they have the ability of hosting active rare earth dopants which leads to a superior efficiency of radioactive transitions. The higher refractive indices make these glasses a brilliant choice for near and mid-infrared (IR) applications. These glasses are easy to prepare in bulk as well as thin film forms. The flexibility of compositions permits the tuning of optical properties which makes them suitable for infrared photonics [23].

14.4 Optical Properties of Chalcogenide Glasses

As compared to oxide glasses, the interatomic bonds of chalcogenide glasses are weaker due to which their optical band gap energy lies in visible or near IR regions. The atoms of chalcogenide glasses are heavier than that of oxide glasses. Their bond vibration energy and phonon energy are lower than oxide glasses which makes possible excellent transparency in infrared region of electromagnetic spectrum. Though the weak bonding and heavy atoms of chalcogenide glasses make them superb for mid- and far-infrared (IR) applications, some other physical properties like glass transition temperature, strength, hardness, etc., are lower than oxide glasses. The high linear refractive indices (2 or 3) of these glasses are due to their high density and polarizability, which mainly depends on the composition of the glasses.

Chalcogenide glasses have been attracting researchers for decades because of their higher Kerr nonlinearity than oxide glasses, like silica [24, 25], which can be derived from electronic processes having ultra-short lifetime (generally less than 50 fs) [26] instead of free electrons having larger lifetime (usually in the orders of nanoseconds). In chalcogenide glasses, nonlinear absorption processes also take place. Multi-photon absorption is one of the examples of nonlinear absorption, in which electron–hole pairs can be created on the absorption of multiple photons. The structural flexibility and the presence of lone pairs in chalcogenide glasses make these glasses suitable to exhibit numerous photo-induced phenomena [25]. When a light, energy of which is equal to the band gap of a chalcogenide glass, is made to fall on the glass, a photo-induced change can be observed in its refractive index which leads to the change in the absorption of the light by the glass [27–30]. There are two mechanisms on which change in refractive indices of glasses is dependent: photodarkening and photo-volume expansion. Photodarkening increases refractive index, while photo-volume expansion mechanism decreases the index.

14.5 Optical Losses in Chalcogenide Glasses

As discussed earlier we know that the atoms of chalcogenide glasses are heavier than the atoms of oxide glasses. Also these glasses are having lower bond vibration energies and phonon energies as compared to oxide glasses. Due to these factors, chalcogenide glasses show outstanding transparency in infrared region of electromagnetic spectra. Investigation of loss mechanism in chalcogenide glasses is very essential for analysing the optical transparency completely.

Chalcogenide glasses are having mainly three optical loss mechanisms, which are as follows: radioactive loss, roughness scattering and material attenuation. Radioactive loss occurs due to coupling into substrate mode caused by bending of waveguide. To reduce this loss, device geometry should be selected properly. The roughness at sidewalls of optical devices causes scattering losses. Material attenuation is classified into two categories: extrinsic loss (occurs due to impurity absorption) and intrinsic loss (occurs due to electronic absorption by band tails).

Over some decades, the third-order optical nonlinearity (TONL) [31]-based devices have been receiving attentions of researchers. The TONL property has been found maximum in chalcogenide glasses as compared to other optical glasses [31] because heavy chalcogen atoms form bonds having low phonon energy. These glasses are amorphous semiconductors, they have been found compatible with fibre telecom as well as silicon technologies, and hence, they can be used in infrared and photonic devices [31]. When chalcogenide glasses are interacted with electrons, X-rays and photons, these glasses can change their phases [32]. This property of these glasses makes them attractive for applications like non-volatile memory devices [32].

14.6 Materials Acceptable for Application and Parameters

Several reports are available on the support of ability of chalcogenide glasses being used in photonic applications. Some of these reports are discussed below:

Samson et al. [33] have reported the fitness of gallium lanthanum sulphide in nanophotonic devices because of its capability to provide switching and modulation functionality. Quemard et al. [34] have reported the studies of Ge–Se–As-based chalcogenide glasses to analyse its nonlinear optical properties. He observed that the nonlinearities of these glasses are 800 times higher than that of silica glasses. These glasses were found appropriate for telecommunication applications. The study on optical and thermal stability of Ge–As–Se-based chalcogenide glasses has been reported by Zhu et al. [35]. The increase of optical and thermal stability in the glasses has been observed with the addition of Ge in the glasses. The glasses are found to be promising for high stable photonic device applications. Dongol et al. [36] have reported the effect of sulphur addition on structural and optical properties of Ge–Se–S-based thin films of chalcogenide glasses. On adding sulphur in the glasses, the optical band gap and the single oscillator energy started increasing, while the refractive index and the dispersion energy started decreasing, which makes the thin films of these glasses a promising applicant for nonlinear optical designs.

The high linear refractive index (which is 2.2–2.6 for sulphide, 2.4–3.0 for selenide and 2.6–3.5 for telluride), low phonon energy and high optical nonlinearity properties of chalcogenide glasses make these glasses a wonderful candidate for supercontinuum applications, reported by Goncalves et al. [37]. These glasses have also been found appropriate in various infrared optical applications as fibres and waveguides because of their ability to extend optical transparency (range is from visible to 10 μm, 12–14 μm and 20 μm for sulphides, selenides and tellurides, respectively) [37]. The other important property of chalcogenide glasses is their high viscoelasticity because of which they can easily be shaped to reduce dimensions forming fibres and thin films that can be incorporated in photonic devices [37]. The optical properties of a chalcogenide glass can be engineered by tuning its compositions.

In GeS2–In2S3–CsCl chalcogenide glasses, the modification of third-order optical nonlinearity with the help of nanocrystallization has been reported by Yang [31]. Nanocrystallization helps to modify the optical band gap in these glasses. With the appearance of nanocrystals in this glass, the nonlinear refraction coefficient as well as the nonlinear absorption coefficient of the glass is increased. The possibility of use of chalcogenide glasses in biomedical applications as sensitive evanescent wave optical sensors was reported by Bureau et al. [38]. They measured the infrared signatures of biomolecules present in human cells. Lezal et al. [23] prepared and investigated the optical properties of As_2Se_3, Ge–Se and Ge–Se–Te-based chalcogenide glasses. The main applications of clalcogenide glasses reported by them are laser power delivery systems, anti-reflection coating, fibre lasers and thermal imaging. The reports on the suitability of Ge–As–Se-based glasses on the manufacturing of buried channel, rib waveguides and nonlinear nanowires were made available by Gai et al. [39]. The production of single mode and multi-mode optical fibres from arsenic sulphide-based chalcogenide glass and their possibility in different applications because of its technical and operating characteristics was reported by Shiryaev et al. [40]. Yang [41] reported the dependency of Ge–As–S chalcogenide glass on its composition in which density of the glass is found to correlate linearly with the sulphur content and the optical band gap is found to decrease with the decrease of content of sulphur in the glass.

Many research works have been done on binary, ternary and quaternary chalcogenide glasses [42–82]. The variations in the optical band gaps and refractive indexes of the glasses have been observed, and the dependence of tuning of their optical parameters on composition has been found. The decrease of optical band gap and the increase of refractive index on adding different elements like As, Ge, In, Sb, Bi, Pb, etc., to most of the chalcogenide glasses have been observed by many researchers [42–82].

14.7 Chalcogenide Glassy Composites Containing Lithium for Application in Photonic Devices

The suitability of lithium containing chalcogenide glasses in photonic applications is reported by many researchers. A few of these reports are summarized below:

Lin et al. [83] reported the fabrication of nano-crystalline IR transparent chalcogenide GeS_2–Ga_2S_3–LiI glass ceramics at temperature 403 °C for different durations. An increase of Li^+ ionic conductivity was observed in the sample of the system heated for the duration less than 60 h. The obtained glass ceramics was found with superior thermo-mechanical properties, enhanced second-order optical nonlinearity and improved ionic conductivity.

Rao et al. [84] reported the rib loading of thin films of lithium nibonate with chalcogenide glassy system $Ge_{23}Sb_7S_{70}$ to fabricate high-performance electro-optical devices (like micro-ring modulators and Mach–Zehnder modulators) on silicon substrate, which can be used for short reach optical interconnects.

Yelisseyev et al. [85] reported that the transparency and photoluminescence of lithium thiogallate $LiGaS_2$ (LGS) are affected by annealing in suitable atmosphere. Suitability of the use of LGS, with anion vacancies, as a tunable light emitting medium in solid-state lasers (because of its nonlinear susceptibility) and in various optoelectronic devices, was also reported by them.

The heterogeneous integration of lithium nibonate and chalcogenide glass waveguides on silicon substrate was reported by Honardoost et al. [86]. This can provide a proficient utilization of second- and third-order nonlinearities on a single chip which makes their suitability in various applications like stabilized optical combs.

Zhang et al. [87] studied the optical properties of $Li_2ZnGeSe_4$ and $Li_2ZnSnSe_4$ semiconductors. The optical band gap shown by these compounds was around 1.8 eV. A large range of their transparency windows was obtained which is 0.7–25 μm. The properties studied by them showed that these compounds are having immense potential to be used in various applications like tunable laser systems.

14.8 Conclusion

In this chapter, glass ceramics and their attractive properties have been discussed. Due to the presence of properties of both glass and ceramic in these materials, they have become popular in many photonic applications like optical amplifiers, telecommunications, spectroscopy, solid-state lasers, light detectors, etc. After that a brief discussion on chalcogenide glasses along with their unique and attractive properties has been given. Chalcogenide glasses have been found as more advantageous than oxide glasses in photonic applications. The use of different suitable materials by different researchers for making various chalcogenide glasses containing lithium ions for photonic applications has also been discussed in this chapter.

References

1. E.D. Zanotto, Am. Ceram. Soc. Bull. **89**(8), 19–27 (2010)
2. Y. Teng, K. Sharafudeen, S. Zhou, J. Qiu, J. Ceram. Soc. Jpn **120**(11) 458–466 (2012)
3. W. Holand, V. Rheinberger, M. Schweiger, Roy. Soc. **361**(1804), 575–589 (2003)
4. B. Boulard, Tran T.T. Van, A. Łukowiak, A. Bouajaj, R.R. Gonçalves, A. Chiappini, A. Chiasera, W. Blanc, A. Duran, S. Turrell, F. Prudenzano, F. Scotognella, R. Ramponi, M. Marciniak, G.C. Righini, M. Ferrari, in *Proceedings Volume 9364, Oxide-based Materials and Devices VI*; 93640Z (2015)
5. D. Dyer, G. Daniel, Oxford Univ. Press **279**, 246–256 (2001)
6. TGP Brochure. www.fireglass.com. Retrieved 24 Nov 2020
7. Our Story. Allegion Corp. Retrieved 24 Nov 2020
8. G.H. Beall, Annu. Rev. Mater. Sci. **22**, 91–119 (1992)
9. T. Rudolph, W. Pannhorst, G. Petzow, J. Non-Cryst Solids **155**, 273–281 (1993)

10. P.W. McMillan, *Glass-Ceramics* (Academic, New York, 1979)
11. E.D. Zanotto, J. Non-Cryst. Solids **219**, 42–48 (1997)
12. P.F. James, J. Non-Cryst. Solids **73**, 517–540 (1985)
13. M.C. Weinberg, J. Non-Cryst. Solids **142**, 126–132 (1992)
14. D.R. Uhlmann, J. Non-Cryst. Solids **41**, 347–357 (1980)
15. I. Gutzow, Contemp. Phys. **21**, 121–137 (1980)
16. W. Vogel, *Glaschemie* (Verlag für Grundstoffindustrie, Leipzig, 1978)
17. T. Komatsu, R. Ihara, T. Honma, Y. Benino, R. Sato, H.G. Kim, T. Fujiwara, J. Am. Ceram. Soc. **90**, 699–705 (2007)
18. W. Höland, G.H. Beall, *Glass-Ceramic Technology* (American Ceramic Society, Westerville, OH, 2001)
19. V.M. Orera, R.I. Merino, BOL SOC ESP CERÁM VIDR **54**(1), 1–10 (2015)
20. P.A. Tick, N.F. Borrelli, L.K. Cornelius, M.A. Newhouse J. Appl. Phys. **78**, 6367–6374 (1995)
21. S. Guddala, G. Alombert-Goget, C. Armellini, A. Chiappini, A. Chiasera, M. Ferrari, M. Mazzola, S. Berneschi, G.C. Righini, E. Moser, B. Boulard, C. Duverger Arfuso, S.N.B. Bhaktha, S. Turrell, D. Narayana Rao, G. Speranza, Glass-ceramic waveguides: fabrication and properties, in *Proceedings of ICTON 2010*, Munich, Germany, paper We. C2.1 (2010)
22. P.A. Tick, Opt. Lett. **23**, 1904–1905 (1998)
23. D. Lezal, J. Pedlikova, J. Zavadila, J. Optoelectron. Adv. Mater. **6**(1), 133–137 (2004)
24. G. Lenz, J. Zimmermann, T. Katsufuji, M.E. Lines, H.Y. Hwang, S. Spalter, R.E. Slusher, S.W. Cheong, J.S. Sanghera, I.D. Aggarwal, Opt. Lett. **25**, 254–256 (2000)
25. A. Zakery, S.R. Elliott, *Optical Nonlinearities in Chalcogenide Glasses and their Applications* (Springer, New York, 2007), p. 135
26. R. Slusher, G. Lenz, J. Hodelin, J.S. Sanghera, L. Shaw, I.D. Aggarwal, J. Opt. Soc. Am. B: Opt. Phys. **21**, 1146–1155 (2004)
27. C. Lopez, Evaluation of the photo-induced structural mechanisms in chalcogenideglass, Ph.D., University of Central Florida (2004)
28. A.V. Popta, R. DeCorby, C. Haugen, T. Robinson, J. McMullin, D. Tonchev, S. Kasap, Opt. Express **10**, 639–644 (2002)
29. G. Yang, H. Jain, A. Ganjoo, D. Zhao, Y. Xu, H. Zeng, G. Chen, Opt. Express **16**, 10565–10571 (2008)
30. P. Lucas, J. Phys. Condens. Matter **18**, 5629–5638 (2006)
31. Y. Yang, T. Sun, C. Lin, S. Dai, X.-H. Zhang, W. Ji, F. Chen, Ceram. Int. **45**, 18767–18771 (2019)
32. N. Ostrovsky, D. Yehuda, S. Tzadka, E. Kassis, S. Joseph, M. Schvartzman, Adv. Optical Mater. 1900652 (2019)
33. Z.L. Sámson, S.C. Yen, K.F. MacDonald, K. Knight, S. Li, D.W. Hewak, D.P. Tsai, N.I. Zheludev, Phys. Status Solidi RRL **4**(10), 274–276 (2010)
34. C. QueÂmarda, F. Smektala, V. Couderc, A. BartheÂleÂmy, J. Lucas, J. Phys. Chem. Solids **62**, 1435–1440 (2001)
35. L. Zhu, D. Yang, L. Wang, J. Zeng, Q. Zhang, M. Xie, P. Zhang, S. Dai, Opt. Mater. **85**, 220–225 (2018)
36. M. Dongol, A.F. Elhady, M.S. Ebied, A.A. Abuelwafa, Opt. Mater. **78**, 266–272 (2018)
37. C. Goncalves, M. Kang, B.U. Sohn, G. Yin, J. Hu, D.T.H. Tan, K. Richardson, Appl. Sci. **8**, 2082 (2018)
38. B. Bureau, X.H. Zhang, F. Smektala, J.-L. Adam, J. Troles, Hong-li Ma, C. Boussard-Ple‘de, J. Luca, P. Lucas, D.L. Coq, M.R. Riley, J.H. Simmons, J. Non-Cryst. Solids **345** & **346**, 276–283 (2004)
39. X. Gai, T. Han, A. Prasad, S. Madden, D. Choi, R. Wang, D. Bulla, B. Luther-Davies, Opt. Express **18**(25), 26635 (2010)
40. V.S. Shiryaev, M.F. Churbanov, Preparation of high-purity chalcogenide glasses, Woodhead Publishing Limited, 2014
41. Y. Yang, Z. Yang, P. Lucas, Y. Wang, Z. Yang, A. Yang, B. Zhang, H. Tao, J. Non-Cryst. Solids **440**, 38–42 (2016)

42. E. Barthélémy, V. Caroline, P. Gilles, B. Marc, A. Pradel, Phys. Status Solid. C **8**, 2890–2894 (2011)
43. R.K. Pan, H.Z. Tao, H.C. Zang, C.G. Lin, T.J. Zhang, X.J. Zhao, J. Non-Cryst. Solids **357**, 2358–2361 (2011)
44. N. Sharma, S. Sharda, V. Sharma, P. Sharma, Mat. Chem. Phys. **136**, 967–972 (2012)
45. M.M. Abdullah, P. Singh, M. Hasmuddin, G. Bhagavannarayana, M.A. Wahab, Scripta Mater. **69**, 381–384 (2013)
46. A.A. Al-Al-Ghamdi, A.K. Shamshad, S. Al-Al-Heniti, F.A. Al-Al-Agel, M. Zulfequar, Curr. Appl. Phys. **11**, 315–320 (2011)
47. Q. Yan, H. Jain, J. Ren, D. Zhao, C. Guorong, J. Phys. Chem. C **115**, 21390–21395 (2011)
48. C.M. Muiva, T.S. Sathiaraj, M.M. Julius, Eur. Phys. J. Appl. Phys. **59**, 10301 (2012)
49. E.R. Shaaban, M. El-El-Hagary, M. Emam, M.B. El-Den, Philos. Mag. **91**, 1679–1692 (2011)
50. S. Sharda, N. Sharma, P. Sharma, V. Sharma, Mater. Chem. Phys. **134**, 158–162 (2012)
51. G. Yang, G. Yann, S. Jean-Christophe, R. Tanguy, C. Boussard-Plédel, T. Johann, L. Pierre, B. Bruno, J. Non-Cryst. Solids **377**, 54–59 (2013)
52. R.K. Pan, H.Z. Tao, J.Z. Wang, J.Y. Wang, H.F. Chu, T.J. Zhang, D.F. Wang, X.J. Zhao, Optik-Int. J. Light Electron Opt. **124**, 4943–4946 (2013)
53. I. Khan, I. Ahmad, D. Zhang, H.A. Rahnamaye Aliabad, S. Jalali Asadabadi, J. Phys. Chem. Solids **74**, 181–188 (2013)
54. K.A. Aly, Appl. Phys. A **99**, 913–919 (2010)
55. A.A. Bahishti, M. Hussain, M. Zuifequar, Radiat. Eff. Defects **166**, 529–536 (2011)
56. R. Mariappan, T. Mahalingam, V. Ponnuswamy, Optik-Int. J. Light Electron Opt. **122**, 2216–2219 (2011)
57. S.K. Sundaram, J.S. McCloy, B.J. Riley, M.K. Murphy, H.A. Qiao, C.F. Windisch, E.D. Walter, J.V. Crum, G. Roman, S. Oleh, J. Am. Ceram. Soc. **95**, 1048–1055 (2012)
58. S.A. Saleh, A. Al-Al-Hajry, H.M. Ali, Phys. Scr. **84**, 015604 (2011)
59. F. Abdel-Wahab, H.A. El Shaikh, R.M. Salem, Physica B **422**, 40–46 (2013)
60. S.S. Fouad, M.S. El-Bana, P. Sharma, V. Sharma, Appl. Phys. A **120**, 137–143 (2015)
61. G. Saffarini, J.M. Saiter, H. Schmitt, Opt. Mater. **29**, 1143–1147 (2007)
62. M. Fadel, S.A. Fayek, M.O. Abou-Helal, M.M. Ibrahim, A.M. Shakra, J. Alloy. Compd. **485**, 604–609 (2009)
63. R. Kumar, D. Sharma, V.S. Rangra, Optoelectron. Adv. Mater.-Rapid Commun. **5**, 1065–1068 (2011)
64. E.R. Shaaban, J. Phys. Chem. Solids **73**, 1131–1135 (2012)
65. F. Chen, W. Yonghui, N. Qiuhua, G. Wang, Y. Chen, S. Xiang, D. Shixun, J. Chin. Ceram. Soc. **41**, 388–391 (2013)
66. R.K. Pan, H.Z. Tao, H.C. Zang, X.J. Zhao, T.J. Zhang, Appl. Phys. A **99**, 889–894 (2010)
67. H.E. Atyia, N.A. Hegab, Physica B **454**, 189–196 (2014)
68. G. Wang, N. Qiuhua, X. Wang, D. Shixun, X. Tiefeng, S. Xiang, Z. Xianghua, Physica B **405**, 4424–4428 (2010)
69. C. Das, M.G. Mahesha, G. Mohan Rao, S. Asokan, Thin Solid Films **520**, 2278–2282 (2012)
70. R. Chauhan, A. Tripathi, A.K. Srivastava, K.K. Srivastava, Chalcogen. Lett. **10**, 63–71 (2013)
71. X. Su, R. Wang, B. Luther-Davies, L. Wang, Appl. Phys. A **113**, 575–581 (2013)
72. R. Naik, A. Jain, R. Ganesan, K.S. Sangunni, Thin Solid Films **520**, 2510–2513 (2012)
73. E.R. Shaaban, Y.A.M. Ismail, H.S. Hassan, J. Non-Cryst. Solids **376**, 61–67 (2013)
74. P. Sharma, M.S. El-Bana, S.S. Fouad, V. Sharma, J. Alloy. Compd. **667**, 204–210 (2016)
75. H.M. Kotb, F.M. Abdel-Rahim, Mater. Sci. Semicond. Process. **38**, 209–217 (2015)
76. N. Sharma, S. Sharma, A. Sarin, R. Kumar, Opt. Mater. **51**, 56–61 (2016)
77. M.M.A. Imran, A.L. Omar, Mater. Chem. Phys. **129**, 1201–1206 (2011)
78. L.K. Abhilashi, P. Sharma, V.S. Rangra, P. Sharma, J. Non-Oxide Glasses **8**, 17–23 (2016)
79. N. Sharma, S. Sharda, S.C. Katyal, V. Sharma, P. Sharma, Electron. Mater. Lett. **10**, 101–106 (2014)
80. S. Sharda, N. Sharma, P. Sharma, V. Sharma, J. Electron. Mater. **42**, 3367–3372 (2013)
81. A. Kumari, A. Sharma, Optik-Int. J. Light Electron Opt. **127**, 48–54 (2016)

82. P. Yadav, A. Sharma, Phase Transit. **88**, 109–120 (2015)
83. C. Lin, L. Calvez, B. Bureau, H. Tao, M. Allix, Z. Hao, V. Seznec, X. Zhang, X. Zhao, Phys. Chem. Chem. Phys. **12**, 3780–3787 (2010)
84. A. Rao, A. Patil, J. Chiles, M. Malinowski, S. Novak, K. Richardson, P. Rabiei, S. Fathpour, Opt. Express **23**, 22746–22752 (2015)
85. A.P. Yelisseyev, M.K. Starikova, V.V. Korolev, L.I. Isaenko, S.I. Lobanov, J. Opt. Soc. Am. B **29**, 5 (2012)
86. A. Honardoost, S. Khan, G. Gonzalez, J. Tremblay, A. Yadav, K.A. Richardson, M.C. Wu, S. Fathpour, in *Proceedings of IEEE Photonics Conference* (2017), pp. 545–546
87. J. Zhang, D. Clark, J. Brant, C. Sinagra III, Y. Kim, J. Jangand, J.A. Aitken, The Royal Society of Chemistry, Dalton Trans. (2015). https://doi.org/10.1039/C5DT01635E

Chapter 15
Battery Applications

Prolay Halder and Sanjib Bhattacharya

Abstract A lithium-ion battery is a type of rechargeable battery, which plays important roles in growing up of electronics device technology. It is used mainly as a portable electronic appliance as a stationary energy storage. In Li-ion batteries, electrolytes are important component for changing and discharging. Several types of electrolytes are used such as the aqueous electrolytes, liquid electrolytes and solid glassy electrolytes. But in Li-ion batteries, the uses of aqueous electrolytes have advantages over liquid electrolytes, but there are many drawbacks such as smaller energy density and limited use due to smaller electrochemical window of stability of 1.023 eV. In case of liquid, electrolytes have high-energy density, no memory effect and low self-discharge, but there are safety problems associated with flammable organic liquid electrolytes. There are many factors for evaluating a good electrolyte material used for lithium-ion battery. Now, the glassy electrolytes and lithium electrodes are suitable choice for Li-ion battery. Glass-based material is a solid-state battery. To improve the safety performance, non-flammable inorganic solid glassy electrolytes are used in the battery. And also the solid lithium-ion battery provides high-energy density in lightweight package. For more improvement for electric vehicles, we have to do further research and need advanced characterization technique.

Keywords All-solid-state lithium-ion batteries · Inorganic lithium-ion conductivity · Glassy electrolytes: advantage and disadvantage

15.1 Introduction

We are essentially concerned about that global warming due to heavy uses of non-renewable energy sources. In search of alternative energy source without pollution

P. Halder
Composite Materials Research Laboratory, UGC-HRDC (Physics), University of North Bengal, Darjeeling 734013, West Bengal, India

S. Bhattacharya (✉)
UGC-HRDC (Physics), University of North Bengal, 734013 Darjeeling, India
e-mail: ddirhrdc@nbu.ac.in; sanjib_ssp@yahoo.co.in

S. Bhattacharya and K. Bhattacharya (eds.), *Lithium Ion Glassy Electrolytes*,
https://doi.org/10.1007/978-981-19-3269-4_15

and also energy-saving storage for portable electronic appliances, we can introduce new technology of lithium-ion rechargeable secondary batteries. The lithium-ion battery industry has occupied a very important position in the world for a long time. Lithium-ion battery is one of the keys to grow up information technology [1]. For a long time, the starter-ignition-lighting (SLI) batteries are used in motor vehicles. The most important application of the rechargeable battery has taken place in portable electronics technology, i.e. laptop, smartphone, tablets, etc. The rapid growth of Li-ion battery (EVs) has become more acceptable by people for vehicles. Therefore, in very short times it expands significantly in market. Most of the economically important things are due to its lightest property, excellent cycle performance and most highly reducing of metals. It confers to battery's good specific density and volumetric energy density [1]. Apart from the large-scale utilization of the rechargeable battery, the Li-ion battery has dominated the maximum part of the fastest going market. The typical value of market share of rechargeable battery chemistries of year 2009 is written as follows [2]: lithium-ion battery: 49%, lead acid: 43%, nickel–metal hybrid (NiMH): 4%, nickel and cadmium (Ni–Cd): 3% and others: 1%.

15.2 Advantages of Li-Doped Glass Composites for Battery Applications

In the recent year, for the domestic uses perspective of lithium-ion battery application, the rapid growth of lithium batteries market share has been increased. In 2018, total global production of lithium-ion battery reached 17.05 GWh, with year-on-year growth of 15.12%. From 2005 to 2018, the global lithium battery market grew from \$5.2 billion–\$35 billion. The annual compound growth rate is 15.1% [3]. Graphical representation of lithium-ion battery market size is shown in Fig. 15.1, and also the same of lithium-ion production is shown in Fig. 15.2.

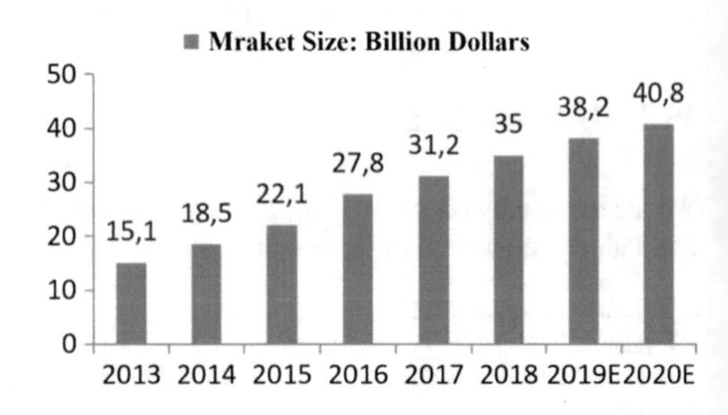

Fig. 15.1 Global lithium-ion battery market size and forecast of 2013–2020. Republished from "Chen Shen and Huaiguo Wang, Journal of Physics: Conference Edition, 2019; IOP Publishing"

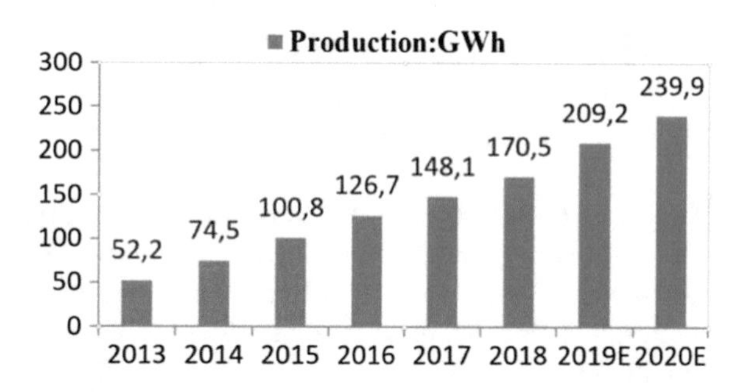

Fig. 15.2 Global lithium-ion battery production and forecast of 2013–2020. Republished from "Chen Shen and Huaiguo Wang, Journal of Physics: Conference Edition, 2019; IOP Publishing".

15.3 Materials Acceptable for Application

Traditional lithium-ion batteries uses organic electrolytes, which are volatile and highly flammable and polymer electrolytes, which have low thermal stability [1]. For this issue, it is easy to cause fire accident and explosion. To avoid this kind difficulty regarding safety issue in lithium-ion battery, liquid electrolytes can be replaced by inorganic non-flammable solid electrolytes [1]. The essential requirements for these solid electrolytes for their purposeful application are as follows:

- Possess high ionic conductivity ($>10^{-4}\ \Omega^{-1}\ cm^{-1}$).
- Negligible electronic conductivity.
- Wide electrochemical stability window as well as favourable transfer impedance between interfaces [4].

Here, two main categories are of great interest, which are crystalline electrolytes and glass-based electrolytes. Several types of solid electrolytes such as perovskite, anti-perovskite, NASICON, garnet, LISICON, sulphide, argyrodite and glass–ceramic are of main focus of researchers for improving cell performance and enhancing electrical conductivity [5].

15.4 Comparison Between Li-Doped and Other Glassy Systems for Battery Applications

There are two types of lithium-ion conductor, one of which is oxide type and another is sulphide-type lithium-ion conductor.

15.4.1 Oxide-Type Conductor

15.4.1.1 Perovskite Conductor

Perovskite (ABO_3)-type conductors have several structural forms. On the basis of optimal bulk Li^+ conductivity with different types of solid electrolytes, A and B sites cation can be replaced with different ions after optimization. From the numerical studies, it is seen that the ionic conductivity can reach 10^{-3} S.cm^{-1} [6]. The general formula of perovskite-type lithium lanthanum titanate can be expressed as $Li_{3x}La_{2/3-x}TiO_3$(LLTO, 0.7 < x < 0.13) [4] with the activation energy ranging from 0.3 to 0.4 eV. LLTO-based solid electrolytes have many importance applications in Li-ion battery uses.

(i) Their appropriate electrical windows and high electrochemical stability (>8 V).
(ii) Acceptable ionic conductivity and negligible electronic conductivity.
(iii) Stability in dry and hydrated atmospheres.
(iv) Stability in wide temperature ranging from 4 to 1600 K [7].

For high gain-boundary resistance and instability against Li metal anode, two major difficulties arise. It has reported that to introduce silica halide, LLZO is acting to reduce the gain-boundary of LLTO and doping of Al to achieve the high conductivity.

15.4.1.2 Anti-Perovskite Conductors

In anti-perovskite (LiRAPs) conductors, the ions in the corresponding lattice sites move electronic inversion with perovskite-type conductor, but the structures are similar to perovskite electrolytes. It is reported by the researcher that at the ambient temperature with activation energy of 0.2–0.3 eV, the ionic conductivity Of LiRAPs can reach greater than 10^{-3} S.cm^{-1}[6], so that even under high temperature, LiRAPs possess good stability. Li^+ with divalent cation, i.e. Mg^{2+}, $Ba^{2+}$and Ca^{2+}, is to be replaced at the Li6O octahedral centre. So that the depletion of LiA can achieve superior conductivity in which the conductivity of $Li_3Ocl_{0.5}Br_{0.5}$ can shift to 6.05 $\times$ 10^{-3} S.cm^{-1} at room temperature [8]. The other type of anti-perovskite can be expressed as $Li_{3-2x}M_xHalO$, where *M* indicates the divalent cations (Mg^{2+},Ca^{2+} or Ba^{2+}) and Hal for halides(cl^-,I^- or mixture). At 25 °C, LiRAPs can show the ultra-fast ionic conduction, and its value is 25 mScm^{-1} [9]. On the basis of these advantages of glassy electrolytes, it is suitable for the application of higher valence dopants for using in future.

15.4.1.3 NASICON Conductor

NASICON is another type of Li-ion conductors with general formula, $Na_{1+x}Zr_2P_{3-x}Si_xO_{12}$ [6]. They are promising fast ion conductors with high conductivity. Reported by the researcher, they are mentioned that there are three reasons in increasing the conductivity of $Na_{1+x}Zr_2P_{3-x}Si_xO_{12}$, and these are to increase the mobility of Na^+ ions in the structure, the higher density of sintered pellets and the Na^+ ions migration through the 3D structure with enlarged tunnel size [10].

The iso-structural form of NASICON-type conductors is $LiM_2(PO_4)_3$ where M stands for Zr, Ti, Hf, Ge and Sn (reported in 1977). They are classified into two ways: the first one is $LiM_2(PO_4)_3$ with rhombohedral symmetry (M = Ti, Ge) and the last one is $LiM_2(PO_4)_3$ with triclinic phase and lower symmetry (M = Zr, Hf and Sn) [6].

Researcher reported that due to the optimal size for conducting Li^+ in $LiTi_2(PO_4)_3$, it has better conductive performance other than $LiM_2(PO_4)_3$, where M stands for Zr, Hf and Sn. And also mentioned that by partial replacing Ti^{4+} cations in $Li_{1+x}R_xTi_2(PO_4)_3$ with trivalent cations such as Al^{3+}, Sc^{3+}, Ga^{3+}, In^{3+} and Cr^{3+} can give higher the conductivity. $Li_{1.3}Al_{0.3}Ti_{1.7}(PO_4)_3$ shows the conductivity of 7×10^{-4} S.cm^{-1} [11].

15.4.1.4 Garnet-Type Oxide

The nominal composition of garnet-type oxides is $Li_5La_3M_2O_{12}$ (M = Tb, Nb) [6], and it shows that lithium-ion conductivity is of the order of 10^{-6} S.cm^{-1} at 25 °C temperature [12]. Ortho silicates garnets possess the general formula $A_3{}^{ii}B_2{}^{iii}(SiO_4)_3$ [6], where A and B are cations, A coordinated with eight oxygen atoms and B coordinated with six oxygen atoms in the structure.

The advantages of garnet-type oxides are that contribution gain boundaries resistance is much smaller than NASICON phosphates and perovskite-type oxides. It also has compatibility with lithium anode. They have high ionic conductibility and excellent chemical stability.

In$Li_5La_3M_2O_{12}$ is expected to enhance the conductivity level by partially replacing of Y or In at the M sites of it. For an example, at 50 °C the increased conductivity is 1.8×10^{-4} S.cm^{-1} [5] in $Li_{5.5}La_3Nb_{1.75}In_{0.25}O_{12}$ with low activation energy. Another type garnet-like structure is cubic $Li_7La_3Zr_2O_{12}$ (LLZO) which has ionic conductivity ~ 3×10^{-4} S.cm^{-1}.

15.4.2 Sulphide-Type Li-Ion Conductors

15.4.2.1 Thio-LISICONs

Due to increase of electrochemical performance as compared with LISICONs family, the tho-LICOCONs have suggested that a sulphide-type Li-ion conductors constitute a family with the nominal composition of $Li_{4-x}M_{1-x}M'_{y}S$, where M stands for AL, Zn and Ga at 25 °C and ionic conductivity of $Li_{4-x}Ge_{1-x}P_{x}S_{4}$ is 2.2×10^{-3} S.cm^{-1} [6]. As the ionic conductivity depends on the ionic radius of the corresponding ions and its polarizability and the interactions between L^{+} and $S^{-2}$are found to be much weaker than the interaction between O^{-2} and L^{+}, by replacing O with S, the interaction between L^{+} in the sub-lattice can be decreased to enhance the concentration of mobile Li^{+} so that the conductivity increases in S-doped LISICON [5].

As crystalline sulphide Li-ion conductor ($Li_{10}GeP_{2}S_{12}$) electrolytes have excellent electrochemical properties, better conductivity and wide range of potential windows, they can be used as an electrolyte in Li-ion batteries, where $LicoO_{2}$ is used as a cathode and Li-In alloy is used as an anode to improve the better cell performance [13].

15.4.2.2 LGPS–Family

$Li_{10}GeP_{2}S_{12}$ (LGPS) was first proposed in 2011 [6]. In solid electrolytes, the Li-ion conduction of $Li_{10}GeP_{2}S_{12}$ type crystalline sulphides is much better than the organic liquid electrolytes, used in lithium-ion batteries. The typical value of ionic conductivity shows over 10^{-4} S.cm^{-1} [5]. However, for the low abundance and high cost of Ge in electrolytes, new type LGPS materials are necessary to introduce for the application of batteries. The LGPS type family is classified into two groups [14]. The first one is Ge-free electrolytes ($Li_{10}GeP_{2}S_{12}$), and the last one is Ge-doped electrolytes ($Li_{7}GeP_{2}S_{8}$) on the basis of this Ge^{4+} being substituted with Sn^{4+} or Si^{4+} such as $Li_{10}SnP_{2}S_{12}$ and $Li_{11}Si_{2}PS_{12}$, respectively. The researchers reported that Sn^{4+} substitution can deliver the bottleneck of Li + diffusion along the z-direction. As a result, $Li_{9.5}Si_{1.74}P_{1.44}S_{11.7}cl_{0.3}$ performs an excellent high conductivity up to 2.5×10^{-2} S.cm^{-1} [15], and also it shows the highest conductivity ever reported for Lithium-ion conductor. $Li_{9.6}P_{3}S_{12}$ has excellent electrochemical stability, and under extreme cell operation condition, it has longer lifespans. The dual doping of Sn and Si with adjusting ratio Sn/Si in original LGPS materials may increase the ionic conductivity level of 1.1×10^{-2} S.cm^{-1} [6].

Table 15.1 Ionic conductivities of representative crystalline inorganic solid Li-ion conductors at room temperature

Electrolyte	Ionic conductivity (S cm^{-1})	References
Li3xLa(213)-xTi03	10^{-3}	[6]
LiRAP	10^{-3}	[6, 8]
LiM2(P04)3	10^{-3}–10^{-4}	[11]
LisLa 3M2012	10^{-4}–10^{-6}	[5]
Li4-xMl-„	10^{-2}–10^{-3}	[6]
LGPS	1.1x 10–10^{-4}	[5, 15, 16]
Li6PSsX	10^{-2}–10^{-3}	[17]

15.4.2.3 Argyrodites

The general formula of Li-ion conducting argyrodites family is expressed as Li_6PS_5X, where *X* stands for halides Cl, Br, I. Li_6PS_5Br shows room temperature ionic conductivities, which applies the process of solid-state sintering [16] in argyrodite. We can express Li_2S sintered with Li_6PS_5Cl and Li_6PS_5Br, which are showing the ionic conductivity of 1.8×10^{-3} $S.cm^{-1}$ and 1.3×10^{-3} $S.cm^{-1}$at room temperature [17]. In several batteries, Li_6PS_5Br is usually employed as a solid electrolytes. The Cu-Li_2S/Li_2PS_5Br/In battery shows high initial charge and discharge capacities of 500 $mAhg^{-1}$ and 445 $mAhg^{-1}$, respectively [18]. At room temperature, the conductivity of crystalline inorganic solid Li-ion conductors is given in Table 15.1.

15.5 Advantages and Disadvantages of Inorganic Lithium-Ion Conductor for Battery Application

Different types of inorganic Li-ion conductors are used as a solid electrolyte in all-solid-state Li-ion battery. They have some advantages and limitations, which are mentioned below:

- Perovskite type of solid electrolytes has high electrochemical stability and stability at high temperature. But they have two major difficulties, its grain boundaries resistance is high and prepares with very harsh condition [6].
- Anti-perovskite types (LiRAPs) of solid electrolytes can reach higher ionic conductivity at ambient temperature. LiRAPs ($Li_{3-x}M_xHal$) can show ultra-fast ionic conductivity in glassy electrolytes and also process good stability. But its disadvantages is that these types of material are attracting and holding water from surrounding. This is the reason why it had prepared in inert atmosphere [4].
- NASICON type of Li-ion conductors in solid electrolytes shows fast ionic conductions, but its grain-boundary resistances are high and the NASICON-type electrolytes solid-state batteries are expensive [19].
- Sulphides-doped Li-ion conductor solid-state electrolytes have much better ionic conductivity compared to oxide-type solid electrolytes in Li-ion batteries. But its

limitation is that this type of material reacts with moisture, so they have instability in ambient atmosphere [20].

- LISICON-type solid electrolytes in solid-state battery have good electrochemical stability and wide potential window voltage. But its ionic conductivity is much lower than the other oxides-type solid electrolytes [5].
- Argyrodite type of solid electrolytes has high ionic conductivity, but they have instability in ambient temperature.

15.6 Parameters for Various Issues

15.6.1 Electrolytes

Lithium-ion battery is a type of rechargeable secondary battery which can be charged and discharged. A solid lithium-ion battery is composed of three major components which are an anode, a cathode and a solid electrolyte.

- Ionic conductivity of the glasses: Lithium conducting glasses exhibit high ionic conductivity than corresponding crystal because of the large amount of free volume of the glasses due to open structure [21]. The structure of SiO_2 glass and crystal is shown in Fig. 15.3.
- Single cation conduction: Inorganic glassy materials produce single cation conduction and which results in very wide electrochemical windows. For the single cation conduction, the inorganic glassy system shows that process in which the mode of ion conduction relaxation is decoupled from the mode of structural relaxation [22]. This system has decoupling property, which generates the higher single ion conduction.

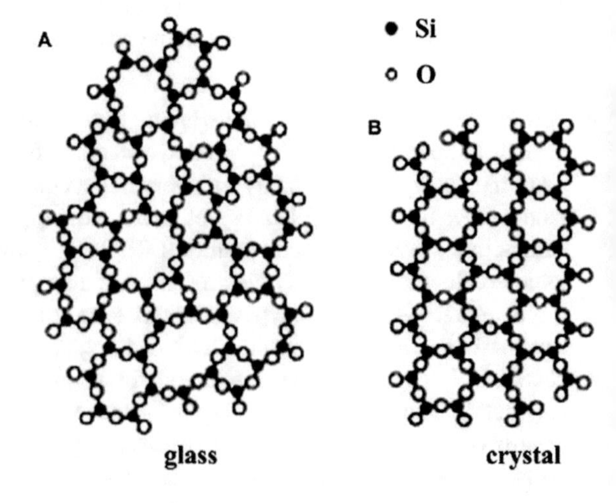

Fig. 15.3 Schematic of structure of sio_2: **a** glass and **b** crystal. Republished from "Can Cao, Zhuo-Bin Li, Xiao-Liang Wang, Frontiers Energy **2** (2014) 1"

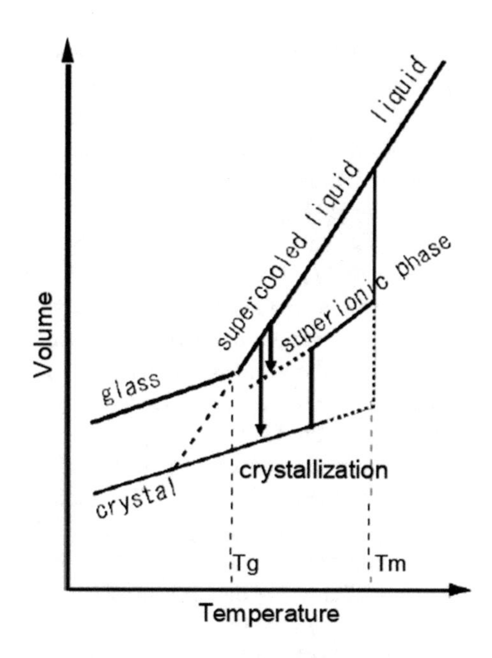

Fig. 15.4 Volume changes from glass to crystal with increasing temperature. Republished from "Courtney Calahoo, Lothar Wondraczek, Journal of Non-Crystalline Solids: X **8** (2020) 100,054"

- Formation of super-ionic metastable phase in glassy materials: Metastable phase is formed by inorganic glassy material [8]. Metastable phase is not thermodynamically stable at a given temperature. If we heat the glass at glass transition temperature, then its conductivity will decrease with crystallization. The variation of volume with temperature is depicted in Fig. 15.4 in the glass-based solid electrolytes using in lithium-ion battery.
- Several ways of increasing conductivity: There are several process for increasing the conductivity of glassy electrodes, i.e. mixed an ion effect, mixed glass former or network former or modifier effect.

At room temperature, Li_2O-B_2O_3 glass electrode increases the ionic conductivity from 1.2×10^{-8} to 8×10^{-7} S cm^{-1} by the addition of network former for modifier SeO_2 [23].

Another way to increase the ionic conductivity of 67Li2S.33P2S_5 glass at room temperature, from 10^{-4} to 10^{-3} S cm^{-1}, is to add 45 mol% of LiI [24].

In order to maximize the conductivity of the glass, electrode is mixed with sulphide and oxide, and this is known as mixed former effect [25].

15.6.2 Various Features

Here, discussion is made about different types of approaches to develop the high ionic conductivity, optimizing electrolyte/electrode interface and better cell performance in solid-state lithium-ion battery for glass–ceramic xLi_2S-$(1-x)P_2S_5$ as a electrolyte [7]. Crystallizing the precursor glass, glass–ceramic electrolytes can be produced. Like the case of glassy electrolytes, we can classify glass–ceramics into oxide and sulphides. Sulphide ion has more ionic radius and high polarizability than oxide ion, so the ionic conductivity of sulphide glass–ceramic is higher than that of oxide glass–ceramic [25].

At room temperature, the conductivity of Li_2S-P_2S_5glass–ceramic can be 10^{-3} S cm^{-1} [7]. It is shown in Fig. 15.5 that optimization condition of heat treatment decreases the gain-boundary resistance and by which the densification process enhances the overall conductivity of a glass–ceramic conductor, but this is not done by enhancing the ionic conductivity in the bulk [26]. By optimized heat treatment, typical value of ionic conductivity of Li_2S-P_2S_5 is found to be 1.7×10^{-2} S.cm^{-1} at room temperature [27].

The crystallization of glass has lower conductivity, but the formation of glass–ceramic and the conductivity will increase by different process like heat treatment and single-step ball milling process [27]. Composition dependency of conductivity of oxide and sulphide glass-based material is shown in Fig. 15.6. It is clear that Li_2S-P_2S_5 glasses enhance the ionic conductivity by heating, and its ionic conductivity is much higher than the precipitated crystalline phases of Li_2S-SiS_2 glasses.

Application of Li_2S-P_2S_5 must arise to all types of solid-state secondary batteries using typical active materials such as $LiCoO_2$ and $Li_4Ti_5O_{12}$ as a positive electrode. Laboratory-scale all-solid-state cell using $LiCoO_2$ as a positive electrode and Li_2S-P_2S_5 glass–ceramic as an electrolyte [28] (In/Li_2S-P_2S_5 glass–ceramic/$LiCoO_2$) is shown in Fig. 15.7.

Basically, this cell is made of three layers. The first layer is a negative electrode (SnS-P_2S_5 glass). The second layer is a solid electrolyte, Li_2S-P_2S_5 glass–ceramics powder. The third layer is a positive electrode which is composed of $L_iC_oO_2$, solid electrolyte (SE) and AB with a weight ratio of 20:30:3. The composite electrode is a mixture of three kinds of these fine powders. This cell provides several advantages

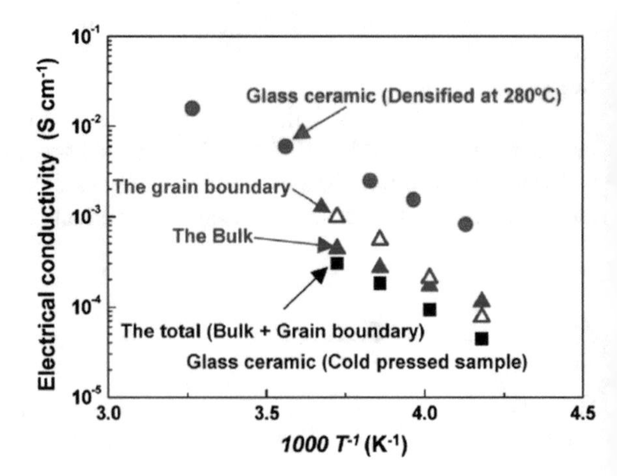

Fig. 15.5 Temperature dependency of the bulk and grain-boundary resistances of the cold-pressed glass–ceramic material. Republished from "Can Cao, Zhuo-Bin Li, Xiao-Liang Wang, Frontiers Energy **2** (2014) 1"

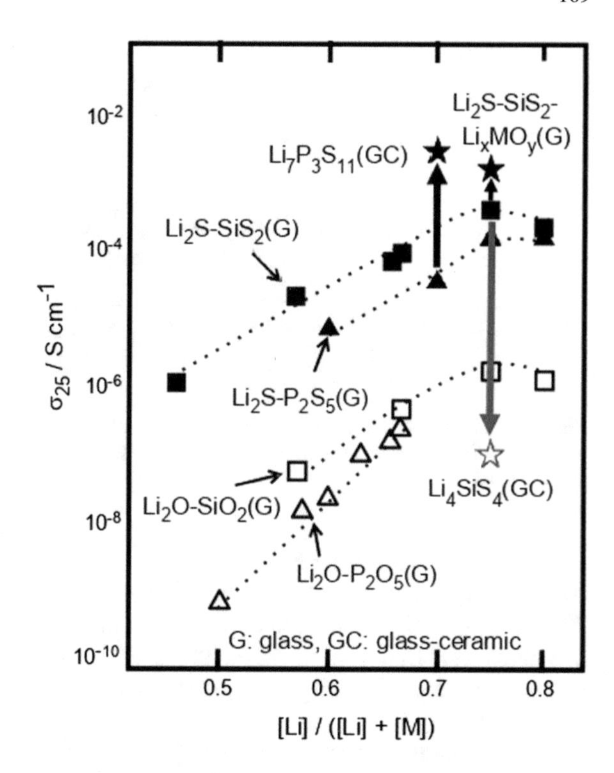

Fig. 15.6 Composition dependence of conductivity at 25°c for oxide and sulphide glass-based materials with high concentration of Li^+ ions. Republished from "Jonathan Lau, Ryan H. DeBlock, Danielle M. Butts, David S. Ashby, Christopher S. Choi, Bruce S. Dunn, Advanced Energy Materials **8** (2018) 1,800,933"

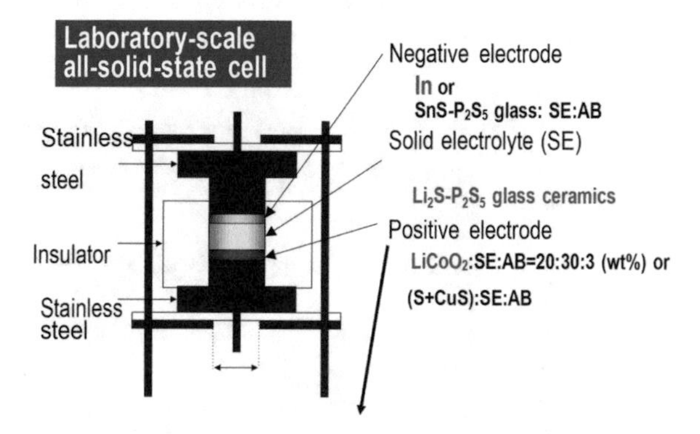

Fig. 15.7 Schematic of laboratory-scale all-solid-state cell, In/Li_2S-P_2S_5glass-Ceramic$L_iC_oO_2$

like it has exhibited excellent cycle performance without loss of any capacity up to 500 cycle numbers [28]. Through solid electrolytes, conducting active materials show the ionic and electronic conduction, which are indicating excellent battery performance.

Laboratory-scale for all-solid-state cell using $Li_4Ti_5O_{12}$ as a positive electrode and Li_2S-P_2S_5 glass–ceramic as an electrolyte (In-Li/70Li_2S·29P_2S_5·1P_2S_3 glass–ceramic/$Li_4Ti_5O_{12}$) is shown in Fig. 15.8.

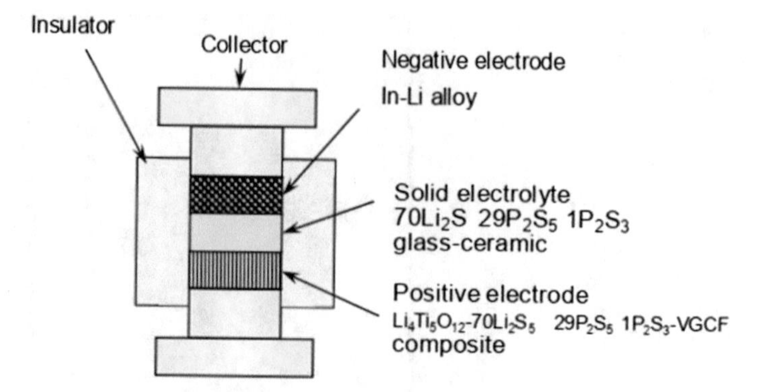

Fig. 15.8 Schematic of a laboratory-scale all-solid-state cell, In-Li/$70Li_2S$ $29P_2S_5$ $1P_2S_3$ glass–ceramic/$Li_4Ti_5O_{12}$. Republished from "Masahiro Tatsumisago, Akitoshi Hayashi, Solid State Ionics **225** (2012) 342"

Basically, this cell is a three-layered system, out of which the first one is a negative electrode (mainly Li-alloy is used), the second layer is solid electrolyte, the $70Li_2S{\cdot}29P_2S_5{\cdot}1P_2S_3$ glass–ceramic powder, and the last one is a positive electrode, a composite of $Li_4Ti_5O_{12}$, $70Li_2S{\cdot}29P_2S_5{\cdot}1P_2S_3$ glass–ceramic and vapour grown carbon fibre [25]. The charge–discharge curves and cell performance of this cell are shown in Fig. 15.9.

As the charge–discharge capacity is 140 m Ah^{-1} without any degradation up to 700 cycle under a voltage-controlled condition between 0.2 and 1.8 at its high-current density, it works reversibility (over 100 $mA.cm^{-2}$). This cell is performed

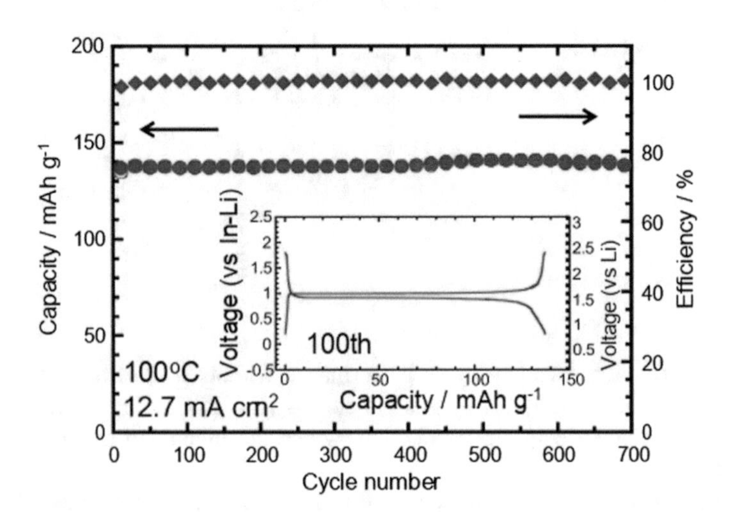

Fig. 15.9 Battery performance of the all-solid-state cell, In-Li/$70Li_2S$ $29P_2S_5$ $1P_2S_3$ glass–ceramic/$Li_4Ti_5O_{12}$. Republished from "Masahiro Tatsumisago, Akitoshi Hayashi, Solid State Ionics **225** (2012) 342"

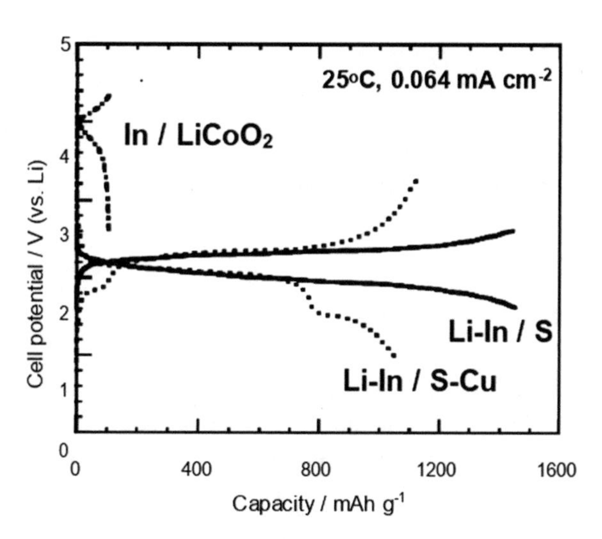

Fig. 15.10 Charge–discharge curves for the all-solid-state cells In–Li/80Li_2S 20P_2S_5 glass–ceramic/S–Cu–SE composites, In–Li/80Li_2S 20P_2S_5 glass–ceramic/S–C–SE composites and In/80Li_2S 20P_2S_5 glass–ceramic/$LiCoO_2$ as the active material. Republished from "Masahiro Tatsumisago, Akitoshi Hayashi, Solid State Ionics **225** (2012) 342"

with negligible volume change during cycling. It exhibits excellent cell performances and long life cycle which is very necessary for solid-state battery.

The charge–discharge curves for all-solid-state cell are shown in Fig. 15.10.

15.6.3 Advancement of Materials

Lithium-ion conducting glassy solid electrolytes, used for all solid-state battery, have certain merits over other such systems. These are mentioned below:

- To improve the safety performance such as to avoid thermal runaway as well as electronic combustion in lithium-ion battery, solid-state glassy materials are used as non-flammable electrolytes instead of liquid electrolytes.
- Glass can be prepared in a wide range of composition. Due to these kinds of characteristics, glass gives batter control over its properties.
- The glassy solid electrolytes have certain advantages over crystalline materials.

  (I) Isotropic conduction properties.
  (II) Absence of grain-boundaries resistance.
  (III) Flexibility of safe and size with satisfactory cost.

- Glassy solid electrolytes have wide range of electrochemical stability in comparison with liquid electrolytes.
- Glassy materials have higher ionic conductivity and higher-energy density.
- Glass–ceramic electrolytes exhibit excellent cell performance and long life cycle.
- The formation of super-ionic metastable phase is the most remarkable advantage of glass–ceramic solid electrolytes.

- In solid-state batteries, in order to make good contract with the electrodes, the glasses need to be grounded into fine powders by mechanical milling techniques. This easy process of synthesis can be performed in room temperature.
- This type of composite materials is easy for film formation.
- In the absence of crystalline pathway in the material, the conduction pathway of sulphide glasses is isotropic.

15.7 Conclusion

Lithium-ion rechargeable batteries have been more attracting for a wide range of uses of portable electronic vehicles (EVs) and PHEVs technology. For using of liquid organic electrolytes and polymer electrolytes in lithium-ion batteries, several safety issues should be realized. To avoid these kinds of thing and improve lithium-ion rechargeable battery, inorganic Li-ion conducting solid materials have been introduced. Inorganic lithium-ion conductors like perovskite-type, anti-perovskite, garnet-type, NASICON-type, thio-LISICONs, LGPS family and glass-based composite material are used in all-solid-state Li-ion rechargeable battery. There are several advantages such as non-flammable safety, reliability and energy density, non-leakage problem rather than liquid organic electrolytes. These may lead to ionic conductivity in the order of 10^{-4}–10^{-2} S.cm^{-1}. Although among these types of Li-ion conductor, the glass-based composite electrolytes are most suitable of all these for solid-state Li-ion battery. Glass-based materials have excellent cell performance. Sulphide-type glassy electrolytes have high ionic conductivity, negligible electronic conductivity, low boundary resistances and wide range of stability. Li_2-S_2P_5 glass–ceramic electrolytes show better charge–discharge capacity with higher cycle number. Li-ion conduction is very smooth between electrode and electrolytes. These are necessary for good battery performance. So glassy electrolytes are the best choice for all-solid-state lithium-ion battery.

References

1. W. Zhang, X. Zhang, Z. Huang, H.-W. Li, M. Gao, H. Pan, Y. Liu, Adv. Energy Sustain. Res. **2100073**, 1 (2021)
2. D.E. Demirocak, S.S. Srinivasan, E.K. Stefanako, Appl. Sci. **7**(7), 731 (2017)
3. C. Shen, H. Wang, Conf. Ser. **1347**, 012087 (2019)
4. F. Zheng, M. Kotobuki, S. Song, M.O. Lai, L. Lu, J. Power Sources **389**, 198 (2018)
5. C. Cao, Z.B. Li, X.L. Wang, X.B. Zhao, W.Q. Han, Frontiers in Energy. Research **2**, 25 (2014)
6. W. Zhao, J. Yi, P. He, H. Zhou, Electrochem. Energy Rev. **2**(4), 574 (2019)
7. O. Bohnke, Solid State Ionics **179**(1–6), 9 (2008)
8. Y. Zhao, L.L. Daemen, J. Am. Chem. Soc. **134**(36), 15042 (2012)
9. M.H. Braga, J.A. Ferreira, V. Stockhausen, J.E. Oliveira, A. El-Azab, J. Mater. Chem. A **2**(15), 5470(2014)
10. H. Aono, E. Sugimoto, Y. Sadaoka, N. Imanaka, G.Y. Adachi, J. Electrochem. Soc. **140**(7), 1827 (1993)

11. R. Kahlaoui, K. Arbi, I. Sobrados, R. Jimenez, J. Sanz, R. Ternane, Inorg. Chem. **56**(3), 1216 (2017)
12. V. Thangadurai, H. Kaack, W.J. Weppner, J. Am. Ceram. Soc. **86**(3), 437 (2003)
13. N. Kamaya, K. Homma, Y. Yamakawa, M. Hirayama, R. Kanno, M. Yonemura, T. Kamiyama, Y. Kato, S. Hama, K. Kawamoto, A. Mitsui, Nat. Mater. **10**(9), 682 (2011)
14. A. Kuhn, V. Duppel, B.V. Lotsch, Energy Environ. Sci. **6**(12), 3548 (2013)
15. Y. Kato, S. Hori, T. Saito, K. Suzuki, M. Hirayama, A. Mitsui, M. Yonemura, H. Iba, R. Kanno, Nat. Energy **1**(4), 1 (2016)
16. H.J. Deiseroth, S.T. Kong, H. Eckert, J. Vannahme, C. Reiner, T. Zaiß, M. Schlosser, Angew. Chem. **120**(4), 767 (2008)
17. H. Schneider, H. Du, T. Kelley, K. Leitner, J. ter Maat, C. Scordilis-Kelley, R. Sanchez-Carrera, I. Kovalev, A. Mudalige, J. Kulisch, M.N. Safont-Sempere, J. Power Sources **366**, 151(2017)
18. M. Chen, R.P. Rao, S. Adams, Solid State Ionics **262,** 83(2014)
19. H. Aono, E. Sugimoto, Y. Sadaoka, N. Imanaka, G.-Y. Adachi, Solid State Ionics **47**, 257 (1991)
20. M. Tatsumisago, A. Hayashi, Int. J. Appl. Glas. Sci. **5**(3), 226 (2014)
21. M. Tatsumisago, Solid State Ionics **175**(1–4), 13 (2004)
22. C.A. Angell, Chem. Rev. **90**(3), 523 (1990)
23. R. Chen, R. Yang, B. Durand, A. Pradel, M. Ribes, Solid State Ionics **53**, 1194 (1992)
24. R. Mercier, J.P. Malugani, B. Fahys, G. Robert, Solid State Ionics **5**, 663(1981)
25. M. Tatsumisago, A. Hayashi, Solid State Ionics **225**, 342 (2012)
26. N. Machida, T. Shigematsu, Chem. Lett. **33**(4), 376 (2004)
27. F. Mizuno, A. Hayashi, K. Tadanaga, M. Tatsumisago, Adv. Mater. **17**(7), 918 (2005)
28. H. Wada, M. Menetrier, A. Levasseur, P. Hagenmuller, Mater. Res. Bull **18**, 189 (1983)
29. A. Hayashi, M. Tatsumisago, Electron. Mater. Lett. **8**(2), 199 (2012)

Chapter 16
Electrochemical Applications

Amartya Acharya, Koyel Bhattacharya, Chandan Kr Ghosh, and Sanjib Bhattacharya

Abstract A lithium-ion battery is a type of rechargeable battery, which plays important roles in growing up of electronics device technology. It is used mainly as a portable electronic appliance as a stationary energy storage. In Li-ion batteries, electrolytes are important components for changing and discharging. Several types of electrolytes are used such as the aqueous electrolytes, liquid electrolytes and solid glassy electrolytes. But in Li-ion batteries, the uses of aqueous electrolytes have advantages over liquid electrolytes, but there are many drawbacks such as smaller energy density and limited use due to smaller electrochemical window of stability of 1.023 eV. In case of liquid, electrolytes have high-energy density, no memory effect and low self-discharge, but there are safety problem associated with flammable organic liquid electrolytes. There are many factors for evaluating a good electrolyte material used for lithium-ion battery. Now, the glassy electrolytes and lithium electrodes are suitable choice for Li-ion battery. Glass-based material is a solid-state battery. To improve the safety performance, non-flammable inorganic solid glassy electrolytes are used in the battery. And also the solid lithium-ion battery provides high-energy density in lightweight package. For more improvement for electric vehicles, we have to do further research and need advanced characterization technique.

Keywords All-solid-state lithium-ion batteries · Inorganic lithium-ion conductivity · Glassy electrolytes: advantage and disadvantage

---

A. Acharya · S. Bhattacharya (✉)
Composite Materials Research Laboratory, UGC-HRDC (Physics), University of North Bengal, Darjeeling 734013, West Bengal, India
e-mail: ddirhrdc@nbu.ac.in; sanjib_ssp@yahoo.co.in

UGC-HRDC (Physics), University of North Bengal, Darjeeling 734013, West Bengal, India

K. Bhattacharya
Department of Physics, Kalipada Ghosh Tarai Mahavidyalaya, Bagdogra, Darjeeling 734014, West Bengal, India

C. K. Ghosh
Department of Electronics and Communication Engineering, Dr. B. C. Roy Engineering College, Durgapur 713026, West Bengal, India

S. Bhattacharya and K. Bhattacharya (eds.), *Lithium Ion Glassy Electrolytes*,
https://doi.org/10.1007/978-981-19-3269-4_16

16.1 Introduction

Electrochemistry is a powerful tool [1] to focus on reactions involving transference of electrons, due to chemical changes. Usually, these chemical changes comprise oxidation or reduction of metal or metal-related complex ions. To understand the difference between a chemical reduction and an electrochemical reduction, consider the example of the reduction of ferrocenium, $[Fe(Cp)_2]^+$ (Cp = cyclopentadienyl), abbreviated as Fc^+, to ferrocene $[Fe(Cp)_2]$, abbreviated as Fc:

- Through a chemical reducing agent: $Fc^+ + [Co(Cp^*)_2] \rightleftharpoons Fc + [Co(Cp^*)_2]^+$
- At an electrode: $Fc^+ + e^- \rightleftharpoons Fc$.

In the first case, electron transfers from $Co(Cp^*)_2$ to Fc^+ are due to difference in energy between the molecular orbitals of the two ions. The transfer of electron in such a process is found to be thermodynamically favourable.

In the second case, Fc^+ is reduced via heterogeneous electron transfer from an electrode. An electrode is an electrical conductor, typically platinum, gold, mercury or glassy carbon. Voltage is applied to the electrode, by means of an external energy source like a potentiostat, to modulate the energy of said electrons. As the electron energies in the electrode become higher than the energies in the LUMO of the Fc^+ ions, the electron gets transferred. Thus similar to the first case, the energy difference of the electrons is the driving force of the process. The difference is changing the driving force of a chemical reduction and requires changing the identity of the molecule used as the reductant [2], which becomes easier in the electrochemical reduction process.

16.2 Materials

A series of glass–ceramics [3], xLi_2O–$(1-x)$ $(0.5ZnO$–$0.5P_2O_5)$ with $x = 0.1$, 0.2, 0.3, 0.4 and 0.5 has been prepared using familiar melt quenching technique with high purity (~ 99%) precursors Li_2O, ZnO and P_2O_5. These chemicals have been weighed properly, and the mixture has been taken in alumina crucible. It was melted in a high-temperature electric muffle furnace in the temperature range 800–850 K depending on the concentration of $Li_2O(x)$. The melts have been equilibrated by stirring to ensure homogeneous mixing. The homogeneous melts have been then instantly quenched at room temperature (300 K) by pouring it in between two aluminium plates. As a consequence, glassy ceramics of thickness ~ 1 mm have been formed.

Glassy ceramics [4], xLi_2O–$(1-x)$ $(0.8V_2O_5$–$0.2ZnO)$ with $x = 0.1$, 0.2 and 0.3 have been developed by solid-state reaction. The precursor powders Li_2O, V_2O_5 and ZnO have been thoroughly mixed in proper stoichiometry of the composition. The mixtures are then heated in an alumina crucible in an electric furnace in the temperature range 600–700 °C for 30 min. Next the mixtures are allowed to pass through the process of slow cooling for 17 h. The final product is gently crushed to get fine powder. Using a pelletizer at a pressure of 90 kg/cm^2, small pellets (diameter ~ 20 mm and thickness ~ 6 mm) of them have been formed.

To check electrochemical stability of the resultant composite, cyclic voltammetry study has been conducted in the 0–2 V window (Model: CHI700E). Here, a three-electrode set-up has been used with a glassy carbon as working electrode, a platinum wire as a counter electrode and a ferrocenium/ferrocene (Fcþ/Fc) as pseudo-reference electrode. While the current flowed between the working and counter electrodes, the reference electrode has been used to accurately measure the applied potential relative to a stable reference reaction. In case of present experiment, the used solvent is Aceto-Nitryl, which has an electrochemical stability of 0–2 V window.

16.3 Properties

16.3.1 Experimental Evidence

The traces of cyclic voltammograms for $x = 0.2$ and 0.4 are shown in Fig. 16.1a, b, respectively. Each trace contains an arrow indicating the direction in which the potential is scanned to record the data. The arrow indicates the beginning and sweep direction of the first segment [5]. It is observed from Fig. 16.1a, b that the apparent oxidation of the samples with $x = 0.2$ and 0.4 starts at about 1.0 V. The subsequent cathodic scan indicates that the oxidation reaction is not observed in the traces of higher scans as no oxidation peaks are observed in the higher cycle. The maximum current of ~ 25 mA in Fig. 16.1a indicates that large amount of sample is oxidized, but maximum current of ~ 9 mA in Fig. 16.1b indicates that a small amount of sample is oxidized. So, it can be concluded that electrochemical stability may be disturbed as the lithium content of the resultant composite decreases.

Cyclic voltammograms of another systems for $x = 0.1$ and 0.3 have been traced out in Fig. 16.2a, b, respectively. Here, arrowheads indicate the direction for scanning of potential to record the data as well as the beginning and sweep direction of the first segment [5]. It is observed from Fig. 16.2a, b that the apparent oxidation of the samples with $x = 0.1$ and 0.3 starts at about 1.0 V. Figure 16.2a clearly indicates that the apparent oxidation of the samples with $x = 0.1$ starts at about 1.0 V, and it exists up to 1.4 V. But apparent oxidation of the samples with $x = 0.3$ starts at about 1.0 V, and it exists up to 1.7 V. The subsequent cathodic scan points out that the oxidation reaction is not observed prominently in the traces of higher scans as any separate oxidation peaks are not observed there. The maximum current of ~ 35 μA in Fig. 16.2a indicates that only a small amount of sample is oxidized for $x = 0.1$, but maximum current of ~ 40 μA in Fig. 16.2b indicates that a large amount of sample is oxidized for x = 0.3. So, it can be concluded that electrochemical stability of the present system may be disturbed as the lithium content of the resultant composite is higher. It is clear from CV studies that the present system shows well electrochemical stability with lower lithium content.

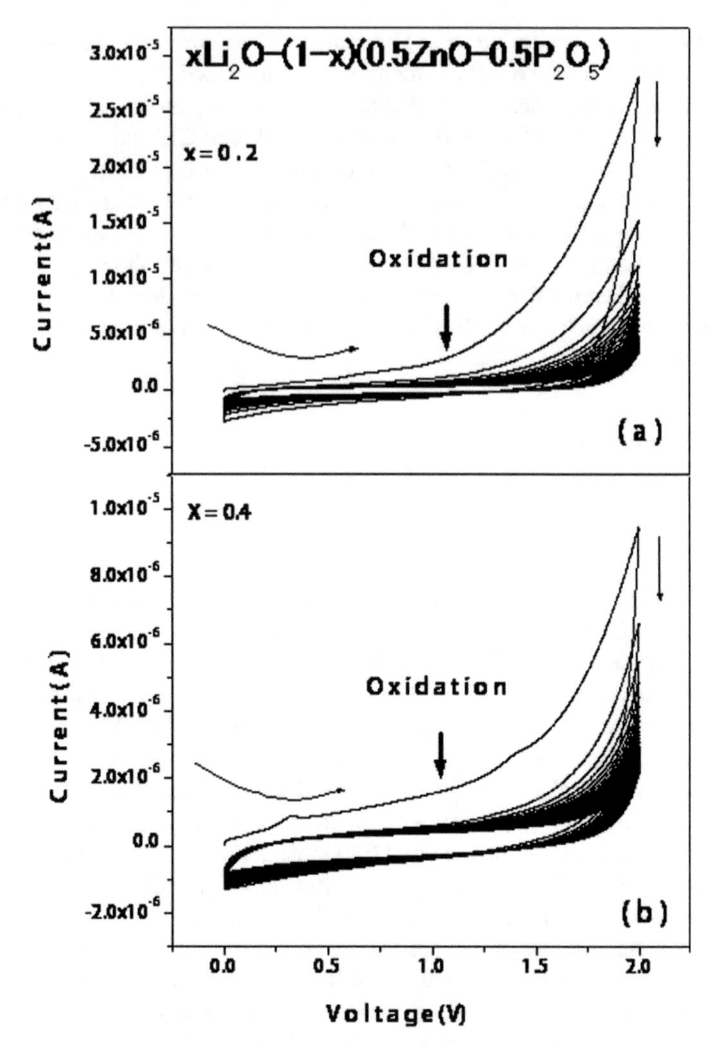

Fig. 16.1 Traces of cyclic voltammograms for **a** $x = 0.2$ and **b** $x = 0.4$, respectively. Republished from "Sanjib Bhattacharya et al, Journal of Alloys and Compounds **786** (2019) 707"

16.3.2 Background Behind Cyclic Voltammetry

16.3.2.1 The Nernst Equation

The Nernst equation relates the potential of an electrochemical cell (E) to the standard potential of a species (E^0) and the relative activities [6] of the oxidized (OX) and reduced (Red) analyte in the system at equilibrium. In the equation, F is Faraday's constant, R is the universal gas constant, n is the number of electrons, and T is the temperature as follows:

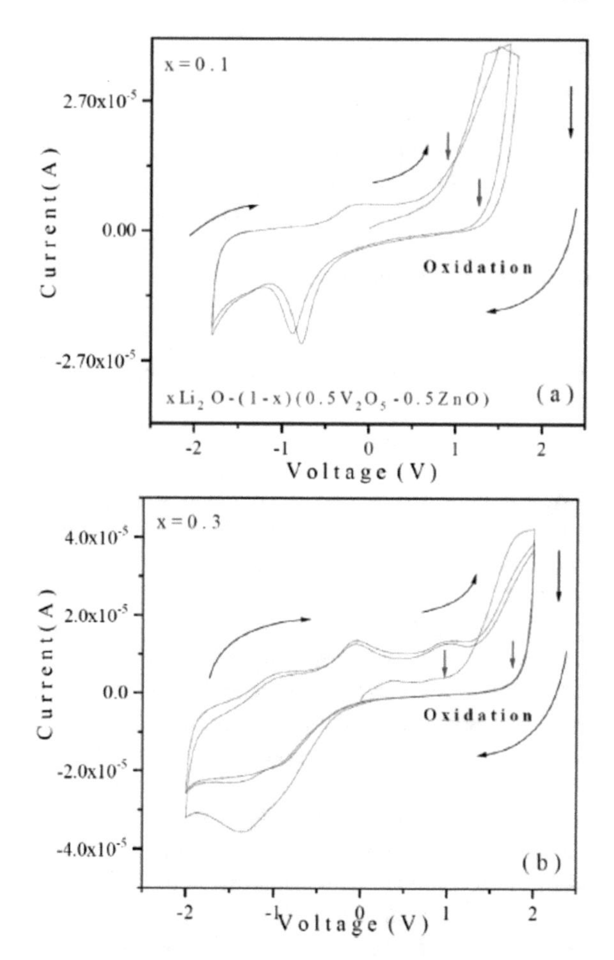

Fig. 16.2 Traces of cyclic voltammograms for **a** $x = 0.1$ and **b** $x = 0.3$, respectively. Down arrow heads indicate oxidation. Republished from "Sanjib Bhattacharya et al, Materials Science and Engineering: B **260** (2020) 114,612"

$$E = E^0 + \frac{RT}{nF} \ln \frac{(OX)}{Red} = E^0 + 2.3026 \frac{RT}{nF} \log_{10} \frac{(OX)}{Red} \tag{16.1}$$

The Nernst equation provides a powerful way to predict how a system will respond to a change of concentration of species in solution or a change in the electrode potential. When a potential is scanned during a CV experiment, the ionic concentration of the solution near the electrode changes with respect to the Nernst equation.

16.3.2.2 Electrolyte Solution

As electron transfer occurs during a CV experiment, electrical neutrality is maintained via migration of ions in solution. As electrons transfer from the electrode to the analyte, ions move in solution to compensate the charge and close the electrical circuit. A salt, called a supporting electrolyte, is dissolved in the solvent to help decrease the solution resistance. The mixture of the solvent and supporting electrolyte is commonly termed the "electrolyte solution".

Large supporting electrolyte concentrations are necessary to increase solution conductivity. As electron transfers occur at the electrodes, the supporting electrolyte will migrate to balance the charge and complete the electrical circuit. The conductivity of the solution is dependent on the concentrations of the dissolved salt. Without the electrolyte available to achieve charge balance, the solution will be resistive to charge transfer. High absolute electrolyte concentrations are thus necessary.

16.3.2.3 Electrode Set-Up (Reference and Counter Electrode)

To conduct the experiment, a three-electrode set-up has been used having a glassy carbon working electrode, a platinum wire used as a counter electrode and a ferrocenium/ferrocene (Fc^{+}/Fc) pseudo-reference electrode. While the current flows between the working and counter electrodes, the reference electrode is used to accurately measure the applied potential relative to a stable reference reaction. A reference electrode has a well-defined and stable equilibrium potential. It is used as a reference point against which the potential of other electrodes can be measured in an electrochemical cell. The applied potential is thus typically reported as "vs" a specific reference. In this case, the reference electrode utilized is also a platinum wire. The reduction potentials should be referenced to an internal reference compound with a known E_0'. Ferrocene is commonly included in all measurements as an internal standard, and researchers are encouraged to reference reported potentials versus the ferrocene couple at 0 V versus Fc + /Fc.

When a potential is applied to the working electrode such that reduction (or oxidation) of the analyte can occur, current begins to flow. The purpose of the counter electrode is to complete the electrical circuit. Current is recorded as electrons flow between the working electrode (W_E) and counter electrode (C_E). To ensure that the kinetics of the reaction occurring at the counter electrode do not inhibit those occurring at the working electrode, the surface area of the counter electrode is greater than the surface area of the working electrode. A platinum wire or disc is typically used as a counter electrode, though carbon-based counter electrodes are also available. When studying a reduction at the working electrode (W_E), an oxidation occurs at the control electrode (C_E).

Applications

Recent work [7] shows a prominent oxidation and reduction peaks of Li_2MnO_3 at 2.98 V and 2.81 V, respectively, with high sweep rate of 500 mV s^{-1}. Li-ion diffusion coefficient value is found to be 1.6×10^{-14} cm^2 s^{-1} and 2.56×10^{-14} cm^2 s^{-1} before and after the cycling, respectively. This Li_2MnO_3 thin films can be utilized as a promising cathode layer in all-solid thin-film battery fabrication.

A high lithium conductive glassy electrolyte has been used to develop the assembly of solid components under isostatic pressure [8]. This cell is advantageous because of their good storage stability and ability to operate until 200 °C. More promising are the thin films solid-state micro-batteries realized by successive depositions of electrodes and electrolyte [xxx]. The low resistance of the electrolyte amorphous layer allows cycling at temperatures as low as −10 °C. This approach indicates the way of commercialization of lithium-ion batteries (LIBs) for the development of portable devices. However, the excessive costs and geographical constraints of lithium resources made it impossible for LIBs to sustain and meet the growing demands of rechargeable batteries.

References

1. K. Takada, J. Power Sources **394** 74e85 (2018)
2. H. Seo, H. Kim, K. Kim, H. Choi, J. Kim, J. Alloy. Comp. **782**, 525 (2019)
3. S. Bhattacharya, A. Acharya, A.S. Das, K. Bhattacharya, C.K. Ghosh, J. Alloy. Compd. **786**, 707 (2019)
4. A. Acharyaa, K. Bhattacharya, C.K. Ghosh, A.N. Biswas, S. Bhattacharya, Mater. Sci. Eng., B **260**, 114612 (2020)
5. N. Elgrishi, K.J. Rountree, B.D. McCarthy, E.S. Rountree, T.T. Eisenhart, J.L. Dempsey, J. Chem. Educ. **95**, 197(2018)
6. X. Yao, B. Huang, J. Yin, G. Peng, Z. Huang, C. Gao, D. Liu, X. Xu, All-solid-state lithium batteries with inorganic solid electrolytes: review of fundamental science. Chin. Phys. B **25**, (2016)
7. V. Paulraj, S. Yasui K.K. Bharathi, Nanotechnol. **32**, (2021)
8. M. Duclot J.L. Souquet, J Power Sources **97–98**, 610(2001)

Chapter 17
Other Applications

Prithik Saha and Sanjib Bhattacharya

Abstract Various glassy samples containing lithium can be used as solid-state batteries of low-power densities due to the low ionic conductivities of solid electrolytes. The nanostructured silicon is promising for high capacity anodes in lithium batteries. The specific capacity of silicon is an order of magnitude higher than that of conventional graphite anodes, but the large volume change of silicon during lithiation and delithiation and the resulting poor cyclability have prevented its commercial application. This challenge could potentially be overcome by silicon nanostructures that can provide facile strain relaxation to prevent electrode pulverization, maintain effective electrical contact, and have the additional benefits of short lithium diffusion distances and enhanced mass transport.

Keywords Solid state battery · Specific capacity · Short lithium diffusion

17.1 Materials

The four most predominant compounds [1–3] formed from lithium reactions are lithium hydride (LiH), lithium oxide (Li_2O), lithium nitride (Li_3N), and lithium hydroxide (LiOH). Table 17.1 shows some properties and discretions of this compound. All are stable but extremely reactive and corrosive compounds.

LiOH: Corrosive; no metal or refractory material can handle molten lithium hydroxide in high concentrations.

Li_2O: Highly reactive with water, carbon dioxide, and refractory compounds.

P. Saha
Department of Electronics and Communication Engineering, Dream Institute of Technology, Kolkata 700104, West Bengal, India

S. Bhattacharya (✉)
UGC-HRDC (Physics), University of North Bengal, Darjeeling 734013, West Bengal, India
e-mail: ddirhrdc@nbu.ac.in; sanjib_ssp@yahoo.co.in

S. Bhattacharya and K. Bhattacharya (eds.), *Lithium Ion Glassy Electrolytes*,
https://doi.org/10.1007/978-981-19-3269-4_17

Table 17.1 Properties of lithium compounds

Formula	LiOH (s)	Li_2O (s)	Li_3N (s)	LiH (S)
Molecular weight	23.95	29.88	34.82	7.95
Density (g/cm3) at 15–20°c	2.54	2.01	1.38	0.78
Melting point (°c)	471.1	1427	840–850 275	688
Boiling point (°C)	925	1527	–	–
ΔG° (kcal/mole) at 25 °C)	−48.99	−133.96	−37.30	−16.72
ΔH° (kcal/mole) at 25 °C	−48.70	−142.65	−47.50	−21.61

Li_3N: Very reactive; no metal or ceramic has been found resistant to molten nitride. The hygroscopic forms ammonia in the presence of water.

LiH: Reduces oxides, chlorides, sulphides readily; reacts with metals and ceramics at high temperatures.

17.2 Experimental

17.2.1 Studies in Lithium Oxide Systems: Lithium Phosphate Compounds (Li_2O-P_2O_5)

In the system Li_2O-P_2O_5, the three compounds Li_3PO_4, $Li_4P_2O_7$, and $LiPO_3$ melt congruently, and therefore, the pyrophosphate exists in two polymorphic forms [4]. This behaviour is analogous to it within the system MgO-P_2O_5 which has three congruently melting ortho-, pyro-, and metaphosphate compounds and a rapid, reversible inversion within the pyrophosphate compound around 68°C[6].The system does not seem to resemble the system Na_2O-P_2O_5 in sight of the inorganic polymorphic behaviour of the sodium phosphate compounds reported by an excellent group of investigators. Chemically pure lithium carbonate and dibasic ammonium orthophosphate were used to prepare the compounds and several compositions between the compound compositions[1]. Some of the mixtures obtained by solid-state reactions were melted in a platinum crucible at about 1000 °C. To minimize the vaporization of the components and the deterioration of the platinum crucible, the melting time was confined to 5 min. After melting, the outside of the crucible was quenched in water. Lithium orthophosphate (Li_3P0_4) was easily formed by solid-state reaction or crystallization from the melt. The congruent melting point was 1225 °C as determined by differential thermal analysis. No glass could be formed by quenching, so the melting point was taken as the temperature at which the charge changed its shape in the quench packet. Lithium pyrophosphate compound did not form a glass on quenching. The congruent melting point was determined by the differential thermal analysis method. Lithium metaphosphate ($LiPO_3$) melted congruently at 688 ± 5 °C as determined by the quenching method. The melting point determined by differential thermal analysis was 665 °C.

17.2.2 *Lithium Oxide Effect on the Thermal and Physical Properties of the Ternary System Glasses (Li_2O_3-B_2O_3-Al_2O_3)*

These glassy systems have been prepared from the subsequent chemical raw materials: orthoboric acid, lithium carbonate, and aluminium oxide [5]. The finely crushed mixture is then placed in a platinum crucible, which is placed to an electric furnace at temperature ranging 1450 °C with a stage for 1 h. The liquid is then cast in an exceedingly graphite mould preheated to approximately 250 °C to limit the thermal shocks during hardening. The samples have been annealed at 250 °C for 1 h. The compositions of studied glasses are given in Table 17.2. To check the system-glass B_2O_3-Al_2O_3-Li_2O_3, five variants are chosen.

The expansion curves of samples have been determined employing a dilatometer DIL 402 °C (Materials Mineral Composite Laboratory (MMCL-Boumerdes-Algeria) at a mean speed of heating of 5 K min^{-1}. The densities are determined out using Archimedes' method with xylene as an immersion fluid. The relative error in these measurements is about $\pm$ 0.03 g cm^{-3}, and also the molar volume V_m is estimated from the molecular weight M and the density ρ according to the relation: $V_m = M/\rho$. Molar volume samples are determined by the following formula: $V_m = P_m$ /MV (molecular weight of glass/density) in mol /cm^3.

By studying this technique, it is noted that with the addition of lithium oxide (10 and 15% Li_2O_3), there is a suppression of phase separation, but at 20% of Li_2O_3 has recurred. Then 25% of Li_2O_3 glass is completely transparent, so the lithium oxide has a great influence on the structural configuration of the glasses studied, and according Li_2O_3/B_2O_3 report, the structure changes. The properties of glassy systems are expected to change according to the structural condition. Generally, good properties are shown in glasses having a high glass transition temperature.

Table 17.2 Chemical compositions of studied glasses (B_2O_3-Al_2O_3-Li_2O_3)

Glass	B_2O_3 (%wt)	Li_2O_3 (%wt)	Al_2O_3 (%wt)
G1	90.00	05.00	05.00
G2	85.00	10.00	05.00
G3	80.00	15.00	05.00
G4	75.00	20.00	05.00
G5	70.00	25.00	05.00

17.2.3 Optical Properties of Lithium Borate Glass $(Li_2O)_x(B_2O_3)_{1-x}$

A series of $(Li_2O)_x(B_2O_3)_{1-x}$has been synthesized with mole fraction x = 0.10, 0.15, 0.20, 0.25, and 0.30 using melt quenching method [4]. The structure of the glass system was determined by FTIR and X-ray diffraction. It is used for detecting penetration of radiation which is applied in homeland security and non-proliferation. The main objectives of the present work were to study the refractive index and optical band gap with variation of lithium borate glass composition.

17.2.4 Characterization and Properties of Lithium Disilicate Glass–Ceramics in the SiO_2-Li_2O-K_2O-Al_2O_3 System for Dental Applications

Glass batches [6] were prepared by mixing appropriate amounts of SiO_2, Li_2CO_3, $MgCO_3$ (Sigma Aldrich Company, Belgium), Al_2O_3 (Fluka Analytical, Germany), P_2O_5 (Acros organics, USA), K_2CO_3 (Fluka chemika, France), and CaF_2 (Merck chemicals, Germany). A two-stage heat treatment schedule was performed in which the glasses were heated to a nucleating temperature of 500 °C with a heating rate of 5 °C/min, held for 2 h, and then ramped up to various crystal growth temperatures (e.g. 700 °C). The heating rate was 5 °C/min, and samples were held for 2 h followed by furnace cooling with 5 °C/min to room temperature. Phase analyses of the glasses and glass–ceramics were performed by X-ray diffraction (XRD, Rigaku TTRAX III) operating from 10° to 70° 2θ at a scan speed of 2° 2θ/min and a step size of 0.02° 2θ with Cu$K\alpha$ radiation ($K\alpha$ = 1.5406 nm) at 300 mA and 50 kV. Identification of phases was achieved by comparing the result diffraction patterns with the ICDD (JCPDS) standard.

The SiO_2-Li_2O-K_2O-Al_2O_3 [6] system was prepared to investigate the effect of glass compositions on their crystal formations, microstructures, and properties through the conventional glass melting process. P_2O_5 and CaF_2 as nucleating agents were introduced to induce heterogeneous nucleation and then produce a fine-grained interlocking microstructure after heat treatment. MgO was added in the glass system to increase the viscous properties, and, finally, the SiO_2: Li_2O ratio in the glass composition was increased. The experimental results and their discussion are addressed as concerns the crystallization behaviour of the glasses, the microstructures, and properties of the glass–ceramics with the potential to be used as dental restorations.

17.2.5 *Properties of Unconventional Lithium Bismuthate Glasses*

Glassy samples of compositions xLi_2O–(100−x) Bi_2O_3 (x = 520, 25, 30, 35 mol %) were prepared using reagent-grade chemicals Bi_2O_3 and Li_2CO_3 [4, 5]. The mixtures of these chemicals taken in alumina crucibles were calcined at 450 °C for 2 h and then melted at 900–1000 °C for 30 min in an electric furnace. Glassy samples were obtained by quenching the melts using two copper plates. All samples were transparent and yellow in colour and showed hygroscopic nature. X-ray diffraction patterns of the as-prepared samples and heat-treated samples at different temperatures for different durations of time were recorded in an X-ray diffractometer ~ Seifert, model XRD 3000 P. The scanning electron micrographs of the polished surfaces of the prepared as well as the heat-treated samples were taken in a scanning electron microscope ~ Hitachi, model S-415A. A thick ~ 150 Å gold coating on the polished surface of the sample was done by vacuum evaporation for the conducting layer function before taking the micrographs. The density of the as-prepared samples was measured at room temperature by the liquid displacement method.

17.2.6 *Effect of Li_2O and Na_2O on Structure and Properties of Glass System (B_2O_3-ZnO)*

Two glass systems were studied [7]: Li_2O–B_2O_3–ZnO (LBZ) and Na_2O–B_2O_3–ZnO (NBZ). The glasses selected were prepared starting from the following chemical raw materials, lithium carbonate, sodium carbonate, orthoboric acid, and zirconium oxide. The finely crushed mixture was then placed in a platinum crucible and introduced to an electric furnace at temperature ranging 1000 °C with a bearing for one (01) h. The liquid was then cast in a graphite mould preheated to approximately 250 °C to limit the thermal shocks during hardening. A decrease in the surface tension of the samples is observed as a result of the addition of the boron oxide B_2O_3 which is known by its influence on the reduction of the surface tension of the glass baths.

17.2.7 *Crystallization Characteristics and Properties of Lithium Germanosilicate Glass–Ceramics Doped with Some Rare Earth Oxides*

The parent glass samples were prepared by the conventional melt quenching technique [8]. Preparation of 40 g powder of glass batches was performed by mixing high purity (purity > 99%) chemical grade of purified quartz SiO2, Li_2CO_3, and GeO_2, with Y_2O_3, La_2O_3, In_2O_3, or CeO_2 powders which were chosen as the raw materials according to the designed molar compositions of the studied glassy systems. Glass–ceramic materials based on the 33.60 Li_2O–66.40 SiO_2 composition of lithium disilicate-$Li_2Si_2O_5$modified by partial GeO_2/SiO_2 replacement (5 mol%) were investigated. Also, the glass–ceramics containing Y_2O_3, La_2O_3 or CeO2 rare earth oxides as well In_2O_3 were successfully prepared and characterized. Varieties of crystalline phases were detected in the developed glass–ceramics through the heat treatment process including lithium di- and meta-silicate solid solutions with GeO_2 [$Li_2(Ge,Si)_2O_5$ and $Li_2(Ge,Si)O_3$], lithium disilicate, di-yttrium disilicate, lanthanum disilicate, lithium indium silicate of pyroxene family, lithium germanate, and cerium dioxide phases. The density of glass–ceramics ranged from 2.38 to 3.14 g/cc, while the microhardness ranged between 4150 and 5585 MPa. The addition of In_2O_3 in the investigated LG glass greatly decreases the leachability of the corresponding crystalline sample (e.g. LGn), i.e. the chemical durability was highly improved. This may be due to the crystallization of the most durable $LiInSi_2O_6$ phase of pyroxene family [8] The obtained materials with such properties are promising for different applications as a new type of electrolyte in solid oxide fuel cells (SOFCs), catalytic converters, or biomedical, as well as dental restorative applications.

17.2.8 Thermal, Mechanical, and Electrical Properties of Lithium Phosphate Glasses Doped with Copper Oxide

Lithium oxide was obtained from lithium carbonate (Li_2CO_3), and phosphorous pentoxide was obtained from ammonium dihydrogen phosphate ($NH_4H_2PO_4$). Copper oxide (CuO) was correspondingly introduced from 10–20 g/100 g of the lithium phosphate glass. Every batch was accurately weighted, thoroughly mixed and added in a porcelain crucible, and melted in an electrical furnace at 1100∘C for 2 h [9, 10]. Rotation for every 30 min was performed for complete homogenization. The melts were cast into preheated stainless steel moulds. The prepared glass samples were immediately transferred to a muffle furnace regulated at 300°C for annealing. The muffle containing the prepared samples after 1 h was switched off and left to cool to room temperature at a rate of 30 C h^{-1}. As the CuO content increased, the colour of the transparent glass changed from light to dark green due to the presence of Cu2 + ions. IR spectroscopy is usually used to study the arrangement of the structural units of the studied glasses. The IR absorption spectra were recorded in the range of 4000–400 cm^{-1} as presented in Fig. 1. Phosphate glasses contain phosphate units which appear in the range of 1400–400 cm^{-1}. H_2O molecules or P–O–H vibrations appear at 3450, 2926, and 2855 cm^{-1} [9, 10]. It is important to highlight that the low-frequency region had no measurable effect on the activation energy of AC conductivity. On other hand, a further increase in the frequency led to a considerable decrease in their values.

17.3 Properties

17.3.1 Lithium Oxide Effect on the Thermal and Physical Properties of the Ternary System Glasses (Li_2O_3-B_2O_3-Al_2O_3)

The borate glasses are known by their structural units made of triangles and tetrahedrons boron in different configurations depending on the percentage of B_2O_3 in the glass chemical composition. The boric anhydride B_2O_3 is sometimes used in many applications such as improving the fusibility; increasing the mechanical resistance, high thermal resistance; a decrease in the surface tension and increases the chemical resistances [3, 4]. For this, the work aim is to study the ternary glass system B_2O_3-Al_2O_3-Li_2O_3, after its preparation and determination of some properties depending on the chemical composition (influence of composition chemical on the properties of glass) and its structure.

17.3.1.1 Physical Properties

It is noted that the density of the samples increased with the addition of lithium oxide, which is explained by filling the voids between the structural units (silica tetrahedra, boron, boron triangles) by the network modifiers (Ion of Li^+), thus decrease the molar volume and density increase [4].

17.3.2 Optical Properties of Lithium Borate Glass $(Li_2O)_x(B_2O_3)_{1-x}$

The refractive index increases with decreasing molar volume and which in turn increases the density [4, 5]. The coordination number of lithium borate glass also leads to the increase of refractive index. An addition of Li_2O causes changed in coordination number and creates more non-bridging oxygen. Thus, it has a higher average coordination number of studied glass which results in the increase of the refractive index. The formation of non-bridging oxygen forms more ionic bonds, which manifest themselves in a large polarizability, thus results in a higher index value. The molar volume decreased as a result from Li_2O occupied interstitial position in the network. The formation of non-bridging oxygen increased with Li_2O resulted in the decreased of optical energy band gap for both direct and indirect band gap. This is because non-bridging oxygen binds excited electrons less tightly than bridging oxygen. The decreased in Urbach Energy is owing to the decrease of degree of disorder in glass structure with Li_2O.

17.3.3 Characterization and Properties of Lithium Disilicate Glass–Ceramics in the SiO_2-Li_2O-K_2O-Al_2O_3 System for Dental Applications

Lithium disilicate glass ceramic ($Li_2Si_2O_5$) is one such all-ceramic system, currently used in the fabrication of single and multiunit dental restorations mainly for dental crowns, bridges, and veneers because of its colour being similar to natural teeth and its excellent mechanical properties [2]. Mechanical properties: the IFT values of the glass–ceramics LD1–LD4 at different temperatures; XRD pattern of LD3 indicated the more numbers of lithium aluminium silicate: virgilite crystals with a high intensity compared to that of other glasses, particularly at 800 °C and 850 °C. Therefore, the thermal expansion mismatch between $Li_2Si_2O_5$ and virgilite resulted in residual stresses or microcracks on cooling, which create crack tip shielding and enhanced toughness of LD3 at 800 °C.

17.3.4 Properties of Unconventional Lithium Bismuthate Glasses

Unconventional bismuthate glasses containing lithium oxide have been prepared by a conventional melt quench technique. X-ray diffraction, scanning electron microscopy, and differential thermal analysis show that stable binary glasses of composition xLi_2O–$(100-x)Bi_2O_3$ can be achieved for $x = 20$–35 mol% [4, 5]. Systematic variation of the glass transition temperature, density, and molar volume observed in these glasses indicates no significant structural change with composition. Differential thermal analysis and optical studies show that the strength of the glass network decreases with the increase of Li_2O content in the glass matrix with a small deviation for the extra stable $30Li_2O$-$70Bi_2O_3$ glass composition. Electrical conductivity: The AC loss and dielectric value of all the glass samples have been measured in the wide frequency range at high temperatures. A high dielectric constant has been observed for all glass compositions similar to copper bismuthate glasses [4–7] which may be due to the high polarizability of the unconventional network former Bi_2O_3. The DC conductivity of the samples has been estimated from an AC complex impedance plot (Table 17.3).

Table 17.3 Crystalline phases developed and properties of the prepared glass–ceramics

Sample	Heat treatment (◦C/h)	Crystalline phases developed	Density (g/cm3)	Hardness (MPa)	Weight loss % (g)
LS	480/5–615/10	Li2Si2O5 (LS2)	2.38	4150	2.05
LG	475/5–600/10	LS2 ss [Li2 (Ge Si)2O5]	2.45	4455	1.85
LGy	530/5–785/10	Li2 (Ge Si)O3, Y2Si2O7	2.88	4670	1.19
LGl	510/5–775/10	La2Si2O7, Li2 (Ge Si)O3	3.14	5585	1.31
LGn	500/5–750/10	Li2 (Ge Si)O3, LiInSi2O6	3.08	4560	0.98
LGc	485/5–615/10	Li2Si2O5, Li6Ge8O19, CeO2	2.51	5050	1.06

17.3.5 Thermal, Mechanical, and Electrical Properties of Lithium Phosphate Glasses Doped with Copper Oxide

Due to the numerous unique properties of phosphate glasses, they have diverse technological applications [1]. Phosphate glasses including copper have special interest due to their several properties. Optical, magnetic, and electrical properties of copper-doped phosphate glasses permit a wide range of applications in solid-state lasers, super-ionic conductors, colour filters, and radiation sensors [2]. Lithium phosphate glass has low melting and low glass transition temperature. It also has high thermal expansion coefficient and high electrical conductivity which qualify glass to be used in laser host matrices, lithium micro-batteries, and electro-optical systems [2, 3]. Many authors had studied the electrical properties of modified phosphate glasses [2, 3]. It has been noticed that ionic conductivity can be reached in many glass systems when different oxides are substituted or added instead of network formers. Also, ionic conductivity increases when any ionic conducting salt or alkali oxide is drugged in the glass matrix [11, 12]. The attention to the ion conduction of phosphate glasses is developed due to many properties of these glasses like high thermal expansion coefficient, low glass transition, and softening temperatures [13, 14]. In order to explore the abilities of phosphate glasses to be successfully applicable in various industrial fields, their properties should be tailored via modifying the composition and inserting beneficial elements within the amorphous network [2, 3]. One promising way to modify the composition of phosphate glasses is by doping with metal oxides. Generally, such doping with metal oxides breaks P–O–P linkages and consequently creates non-bridging oxygen atom. It is extremely important to note that the ionic cross-linking between the non-bridging oxygens of two phosphate chains is granted by modifying cations, which effectively improve the mechanical strength and chemical durability. It is well known that the conductivity of the materials is greatly affected by their structure. Both mobile ions and their concentrations are expected to play an increasingly important role in the occurrence of this effect. It is important to highlight that an increase in temperature and the concentration of mobile ions is responsible for more enhancement of the conductivity of the material [15]. Therefore, DC conductivity of the $50Li_2O - 50P_2O_5$ glass sample with different CuO contents was measured at different temperatures. FTIR showed different absorption bands due to phosphate groups and copper oxide. XRD confirmed the amorphicity of the prepared samples. Both thermal expansion and mass density increased with increasing the addition of CuO. The formation of P–O–Cu bonds increased the cross-link density in the glass network and therefore increased the glass transition temperature (Tg). Based on the variations in the Tg via structural modification, the glasses revealed to be basically ionic conductors. The mechanical properties of glass samples increased with an increase in CuO. Electrical conductivity in contrast to activation energy significantly increased with increasing the CuO content. These increases may be attributed to the possibility of ionic contribution to the electrical conductivity.

17.4 Applications:

17.4.1 Progress in Solid Electrolytes Towards Realizing Solid-State Lithium Batteries

Most of the innovations have been made after the emergence and spread of lithium-ion batteries, as overviewed in this paper [3]. Studies on solid-state batteries had been focused on the enhancement of ionic transport in solid electrolytes in the twentieth century, because solid-state batteries were suffering from the low-power densities due to the low ionic conductivities of solid electrolytes. Development of solid electrolytes with the conductivities of the order of 10^{-3} S cm^{-1} has chased the ionic transport in bulk out of the rate-determining steps to leave the transport at interfaces still highly resistive, which is a new topic in the development of solid-state lithium batteries. The high resistance at the cathode interface in sulphide-electrolyte systems had been the last hurdle for practical power densities and has been overcome by a unique interface design. On the other hand, high grain-boundary resistance in oxide-electrolyte systems is still left, and its reduction is a big challenge in realizing chemically stable solid-state batteries.

17.4.2 Charge Carrier Transport and Electrochemical Stability of Li_2O-Doped Glassy Ceramics

New Li_2O-doped glassy ceramics [13] have been prepared using melt quenching route, and their electrical conductivity has been studied in wide frequency and temperature regime. It is revealed from AC conductivity data that mixed conduction process is responsible for their electrical transport. It is anticipated from the nature of composition that Li + conduction mostly contributes to electrical conductivity at high temperature, and Mott's variable range hopping (VRH) model has been utilized to analyse low-temperature DC conductivity data due to polar on hopping. The traces of cyclic voltammograms reveal the existence of oxidation and the nature of electrochemical stability of the present system, which is expected to be good candidates for lithium-ion battery application.

17.4.3 PH Sensors with Lithium Lanthanum Titanate Sensitive Material: Applications in Food Industry

The lithium lanthanum titanate ceramics [14] as sensitive material for pH sensors have been carried out. The high lithium conductivity of this oxide, at room temperature, would indicate a possible use of this material as a lithium-ion-selective electrode. However, a strong interference in aqueous medium leads to a linear response of the electrode potential as a function of the pH of the solutions and not as a function of the Liþ concentration. Two electrode configurations have been used: a membrane one with an aqueous internal reference and an "all-solid-state" one with metallic titanium internal reference. Linear relationships E = ref ¼ a pH þ b are found although sub-Nerstian responses are observed in both cases. This sensor can be used in industrial processes like milk fermentation, in situ control of cleaning of fermentors (cleaning in place), yoghurt fabrication, and wastewater treatments for examples. The sensor is not sensitive to the variations of the redox potential of the solutions. A chemical stability of the ceramic at high temperature (up to 1000 °C) and high pressure and a good mechanical resistance makes these ceramics valuable for the use as sensitive materials in industrial pH sensors.

17.4.4 Co_3O_4 Nanomaterials in Lithium-Ion Batteries and Gas Sensor

The cyclic voltammograms (CVs) of electrodes [15] made from **Co_3O_4**nanotubes, nanorods, and nanoparticles at a scan rate of 0.5 mVs^{-1} and a temperature of 20 °C exhibit excellent sensitivity to H_2 and alcohol.

17.4.5 Nanostructured Silicon for High Capacity Lithium Battery Anodes

The nanostructured silicon [11] is promising for high capacity anodes in lithium batteries. The specific capacity of silicon is an order of magnitude higher than that of conventional graphite anodes, but the large volume change of silicon during lithiation and delithiation and the resulting poor cyclability have prevented its commercial application. This challenge could potentially be overcome by silicon nanostructures that can provide facile strain relaxation to prevent electrode pulverization, maintain effective electrical contact, and have the additional benefits of short lithium diffusion distances and enhanced mass transport.

17.4.6 Dielectric Studies of Silver-Doped Lithium Tellurite Borate Glasses for Fast Ionic Battery Applications

Frequency-dependent AC conductivity measurements were implied on the prepared glass system in the high-frequency range 5 Hz–35 MHz [12]. AC conductivity measurements were recorded from room temperature to 350 °C for all the samples. Presence of two ionic network modifiers such as Ag_2O and Li_2O plays a significant role in the electrical transport properties like real dielectric permittivity(ε /) and loss tangent (tan δ). Addition of Ag_2O causes breaking of base glass former B2O3 structural units with the creation of non-bridging oxygens (NBOs) in the glass network. Dielectric properties of the glass system show inconsistency in the results from room temperature to 100 °C due to non-uniform charge carrier moments up to certain temperatures. Tangent loss and dielectric permittivity (ε /) decreased with increase of frequency and getting saturation at higher frequencies. Lithium containing glasses show higher-order electrical conductivity, i.e. ionic conductivity. Lithium containing glasses are widely used for electrochemical devices like solid-state batteries, glass electrolytes, and fuel cells. Apart from these applications, Li2O containing glasses are also used for various industrial applications as well as in the formation of dielectric materials for high-speed transmission of signals.

References

1. M.M. Markowitz, Chem. Educ. **40**(12), 633 (1963)
2. R. Roy, E.T. Middleswarth, F.A. Hummel, Am. Mineralogist **33**, 458 (1948)
3. Progress in solid electrolytes toward realizing solid-state lithium batteries, ed. by K. Takada, National institute for materials science, vol. 1–1, (Namiki, Tsukuba, Ibaraki, Japan), pp.305–0044
4. D. Aboutaleb, B. Safi, J. Chem. Mol. Nucl. Mater. Metall. Eng **9** 200(2015)
5. J.E. Shelby, *Introduction to Glass* (The royal society of chemistry, Science and technologies. Immiscibility/Phase separation, 1997), pp. 48–67

6. Characterisation and properties of lithium disilicate glass ceramics in the SiO_2-Li_2O-K_2O-Al_2O_3 system for dental applications, ed by N. Monmaturapoj, P. Lawita, W.Thepsuwan. National metal and materials technology center, vol. 114 (Thailand Science Park, Pathumthani 2013), p. 12120
7. D. Aboutaleb, B. Safi, S. Laichaoui, Z. Lemou, IJMMM **6**, 365 (2018)
8. S.M. Salman, S.N. Salama, Ebrahim, A. Mahdy, bol etín de la soc i edad e s pañola de c e rámi ca y vidr io 58, 94(2019)
9. R.A. Youness, M.A. Taha, A.A. El-Kheshen, N. El-Faramawy, M.A. Ibrahim, Mater. Res. Express **6**, 075212(2019)
10. H.E. Batal, Z.E. Mandouh, H. Zayed, S. Marzouk, G. Elkomyand A. Hosny, J. Mol. Struct. **57**, 1054(2013)
11. R.S. Jeannineand, J. Song, Energy Environ. Sci., **4**, 56(2011)
12. P. Naresh, V. Sunitha, A. Padmaja, P. Uma, N. Narsimlu, M. Srinivas, N.P. Rajeshand, K.S. Kumar, in *AIP Conference Proceedings*, vol. 2244 (2020), p. 100002
13. S. Bhattacharya, A. Acharya, A.S. Das, K. Bhattacharya, C.K. Ghosh, J. Alloys. Compd. **786**, 707 (2019)
14. Cl. Bohnke, H. Duroy, J.L. Fourquett, Sens. Actuators B **89**, 240(2003)
15. W.Y. Li, L.N. Xu, J. Chen, Adv. Func. Mater. **15**, 851 (2005)

Printed in the United States
by Baker & Taylor Publisher Services